CLINICS IN DEVELOPMENTAL MEDICINE No. 125
MUSCLES, MASSES AND MOTION. THE PHYSIOLOGY OF NORMALITY, HYPOTONICITY, SPASTICITY AND RIGIDITY

Clinics in Developmental Medicine No. 125

MUSCLES, MASSES AND MOTION.
THE PHYSIOLOGY OF NORMALITY, HYPOTONICITY, SPASTICITY AND RIGIDITY

E. GEOFFREY WALSH, MD, FRCP, FRSE

1992
Mac Keith Press

Distributed by:
OXFORD: Blackwell Scientific Publications Ltd
NEW YORK: Cambridge University Press

©1992 Mac Keith Press
5a Netherhall Gardens, London NW3 5RN

All rights reserved. No part of this publication may be reproduced, stored in a retrieval system, or transmitted in any form or by any means, electronic, mechanical, photocopying, recording or otherwise, without the prior permission of the publishers

First published 1992

British Library Cataloguing in Publication Data. A catalogue record for this book is available from the British Library.

ISBN   0 521 43229 4

Printed in Great Britain at The Lavenham Press Ltd, Lavenham, Suffolk
Mac Keith Press is supported by **The Spastics Society, London, England**

CONTENTS

FOREWORD	*page* vii
INTRODUCTION	ix
1. CONCEPTS, CLINICAL METHODS AND CONTRACTIONS	1
2. HISTORICAL AND ELEMENTARY METHODS OF MEASURING MUSCLE TONE	22
3. DISPLACEMENT AS THE INDEPENDENT VARIABLE	38
4. TORQUE AS THE INDEPENDENT VARIABLE	43
5. FEEDBACK-INDUCED CHANGES	59
6. MEASUREMENTS OF STIFFNESS AND INERTIA IN ABSOLUTE UNITS	67
7. THIXOTROPY: A TIME-DEPENDENT STIFFNESS	78
8. HYPOTONIA	103
9. THE USE OF ABRUPTLY ALTERNATING TORQUES TO INVESTIGATE THE POSTURAL SYSTEM	118
10. HYPERTONIA: RIGIDITY, SPASTICITY AND OTHER CONDITIONS	133
11. WHOLE BODY VIBRATIONS AND INTEGRATIVE ASPECTS	166
12. GLOSSARY OF BIOMECHANICAL TERMS	191
REFERENCES	205
INDEX	215

To Penelope, without whose
support this book would
never have been started,
let alone finished

FOREWORD

Motor disorders in children, of which cerebral palsy is the principal one, are a heterogeneous group of conditions. They are mainly caused by damage to the central nervous system. However, the effects are seen at the periphery in disturbed function of the muscular system. We can observe that the dysfunctioning system can lead to secondary effects of contractures or abnormal muscle stiffness. A lot of theoretical work in trying to understand these conditions has looked at neurology and the neurophysiology of muscle movement, but often ignored is one vital aspect of the problem. That is that the periphery biomechanical transformation of muscle is an important cause of deformity and easily confused with spasticity. We need to learn more about inertia, resonant frequency, thixotropy, and plastic characteristics of muscle and methods of measuring them, and it was with this in mind that we asked Dr Walsh to present this study of the features of the musculoskeletal system.

Two of the basic deficits that we see are weakness and stiffness. This book is a collection of one man's experiments into the measurement of stiffness at a joint from a physiological and engineering point of view. It is not a neurophysiological treatise, and readers will look in vain for discussions of high-gain cords versus increased sensitivity of spindles or alpha-gamma linkages. What it does provide us with is the absolutely essential basic knowledge of the biomechanics of the muscular system, which we must have if we are to interpret at all the more familiar neurological and neurophysiological investigations of the disturbed motor system.

Geoffrey Walsh is the author of a well-known textbook of physiology[1]; what is less known is his lifelong interest in the study of biomechanics. One of us (J.K.B.) discovered this interest many years ago when as a young paediatric neurologist in Edinburgh he found, along a winding corridor, a sub-basement in the old Medical School where, surrounded by equipment from army surplus stores, resided Geoffrey, then Reader in Physiology, and his man George. J.K.B. soon learned to 'phone rather than visit in order to avoid becoming part of one of their experiments —this after having once been suspended bodily from the ceiling and studied as a pendulum! Dr Walsh has been known to drive around Edinburgh in a home-made steam car, is a painter, saxophonist and an amateur radio enthusiast. He designed what he called his Pugnatron, a totally mechanical load-seeking device which will paint a surface, showing how the mechanics of a system are as important as the electrics. He belongs to the old school of physiology, trained in Oxford and Harvard (a student of Liddell, who was himself a student of Sherrington), in the days when one thought up projects, made the apparatus and did the experiment oneself. He has waited for aircraft to land to haul out the pilot and measure his tremor rating. Fatigued cyclists have had the control of their motor units monitored while still exhausted, and his current work includes studies on high-inertia ankle-jerk torques in the cerebral palsy field.

Treatments of cerebral palsy, many of which have not undergone scientific evaluation, are sometimes approached on a basis more of religious fervour than of scientific aloofness. This has led to polarisation into different, often opposing camps, such as the two orthopaedic approaches of either operating at all joints at once versus the more conservative multiple small operations approach, or the use of rhizotomy, which is bedevilled by the failure to define and measure spasticity so that one is either for it or against it, and again looking perhaps at economic factors rather than scientific ones in deciding whether to proceed. Scientifically, our progress has been impeded by our lack of knowledge of what we are measuring and how we should measure it. This set of fascinating essays will open many clinicians and therapists to a much broader understanding of how the motor system functions.

M.C.O. BAX, D.M., F.R.C.P.,
Senior Research Fellow,
Community Paediatric Research Unit,
Westminster Children's Hospital,
London.

J. KEITH BROWN, M.B., CH.B., F.R.C.P., D.C.H.,
Consultant,
Department of Paediatric Neurology,
Royal Hospital for Sick Children,
Edinburgh.

REFERENCE

[1]Walsh, E.G. (1964) *Physiology of the Nervous System.* 2nd Edn. London: Longman.

INTRODUCTION

Much of the work of the illustrious neurophysiologist Sir Charles Sherrington was concerned with muscle tone. His observations were on spinal and decerebrate cats, and whilst most of the material remains valid, other aspects—some 50 years after his death at the age of 94—are now decidedly dated. I started my career as a medical student in Oxford in 1940. I remember a striking tutorial in 1942 or thereabouts with D. Whitteridge discussing the work of Gordon Holmes on the control of arm movements in patients with cerebellar injuries (Holmes 1917). The Professor of Physiology at that time was E.G.T. Liddell FRS who, with Sherrington 16 years previously, had described the stretch reflex in cats. My own interest in muscle tension grew when, doing clinical neurology, I encountered the problems associated with increased (and sometimes decreased) muscle tension. As a young medical graduate in the late 1940s, I saw something of P.A. Merton's work on the control of the adductor pollicis in his basement laboratory at the National Hospital for Nervous Diseases, Queen Square, London. This was the basis of the much discussed 'servo' theory of muscle control. Although the theory is now partially discredited, the idea led to a most interesting and revealing set of observations. I have had some interest in neuromuscular control for a long time: I published a study on the control of breathing in 1947 and a paper on physiological tremor in 1956. Work on tremor has continued intermittently to the present time, and certain investigations are summarised in this volume.

Sherrington pointed to the importance of the muscle spindles in the regulation of muscle tone. Since his death there has been a great deal of detailed information about these endings and the reader is referred to two important monographs (Matthews 1972, Barnes and Gladden 1985). The role of the stretch reflex may have been overemphasised in the past. The great intricacy of muscle spindle organisation is now apparent. Collectively the muscle spindles represent a sensory system of complexity equal to that of the eye. Whilst an abundance of information is available about peripheral mechanisms, the functional significance of this detailed organisation has largely been elusive.

The use of the word 'tone' in connection with muscle goes back at least to the time of Galen (see Chapter 1). In the past, information was obtained by manipulating a limb manually and sensing the resistance. In moving a limb, speed is an important variable: with rapid motion, reflexes may be seen which are absent with slow movements. This rapid and versatile but non-quantitative method can now be supplemented in selected series by machine measurement as discussed in this book. Some information about muscle tone with the limb at rest can be obtained by electromyography but the conversion from these samples of electrical activity to absolute mechanical units is not straightforward. In a muscle which is not changing its length, however, there is a linear relationship between the EMG voltage and the tension which is generated.

Even if there is no reflex activity, there are a number of different forms of the resistance to motion. There is inertia, spring stiffness, viscosity and thixotropy. Apart from inertia, all of these parameters may be expected to change if the muscles become active. Clearly the analysis of muscle tone is a potentially complex problem requiring at least a moderate amount of technology. A global definition of muscle tone is of little analytical value.

Much of this book is concerned with information which can be obtained by moving limbs. The very nature of the questions which may be asked cannot be approached with clinical methods alone.

There is a long custom amongst traditional scientists of building one's own apparatus. For years I have had a lathe and drilling machine in my laboratory. Immediately after the war there was almost no electrophysiological apparatus commercially available or affordable. The academic staff and technicians in the Oxford and Edinburgh Departments of Physiology were in the habit of visiting surplus stores or yards where ex-service electrical apparatus was acquired for a song. Beating swords into ploughshares and spears into pruning hooks, the 'junk' was recycled into research and teaching apparatus. I have undertaken the electronic work for the instruments described. The talented instrument worker and my associate, Mr George Wright, has developed the mechanical side. Whatever readers and other workers may make of the cluster of observations described in this book, the devices described are certainly unique. If I had depended on commercially available equipment none of the work would have been possible.

It would naturally have been possible to include more material dealing with the nerve pathways concerned, but this is readily available in standard physiological texts.

This book is published by the Mac Keith Press. The name commemorates Dr Ronald Mac Keith, a paediatrician at Guy's Hospital in London, who was highly skilled in getting specialists to talk of their own interests. At one of the famous Oxford meetings, Dr Mac Keith started a discussion on the meaning of the term 'muscle tone'. The problem was exhaustively debated but after two hours no firm conclusion had emerged. It had become apparent to me over the years that here was a topic of great interest to neurologists as well as to physiologists. Before the meeting, I had been considering changing my research interests to see what approaches I could make to the understanding of muscle tone. I had not had a training in engineering and decided that I had to immerse myself in practical mechanical work.

My first approach to muscle tone was to subject people to whole body vibration. British Rail, being subject to fierce questioning in Parliament about the swaying of their coaches ('Bouncing down to Crewe'!), asked me to undertake some observations. Special trains were run down the main London line from Edinburgh to Granthouse on the borders for these observations (some of which are summarised in Chapter 11). The problem was brought into the laboratory with the use of the mechanically driven tilt table.

To some extent this book represents one man's attempt to answer Dr Mac Keith's question. The earlier part of Chapter 4 details two approaches before the printed motor became available. The basic question posed has been to ascertain the effects of varying, in a variety of ways, the force applied to the limb—surely a fundamental question. Apart from the occasional use of intramuscular electrodes the methods have been noninvasive and hence it has been possible to use them to investigate a wide variety of problems both in normal subjects and in patients suffering from neurological abnormalities.

ACKNOWLEDGEMENTS

I have to acknowledge much help from my clinical colleagues, a number of whom are co-authors of scientific works mentioned in these pages, and from Dr Martin Lakie, who worked with me for a number of years and is now at the Applied Physiology Research Unit, Birmingham University. Over the years fruitful discussions have been enjoyed with Dr Geoffrey Rushworth and Mr Phillip Harris FRCS, currently Editor-in-Chief of *Paraplegia*, whilst help with computing has been given by Dr John Welford. I am still active in the assessment of children with cerebral palsy, and acknowledge the help of Dr Keith Brown, Dr Robert Minns and Dr J.-P. Lin. Dr Lin and Mr Ian Corry FRCS made many useful comments on the manuscript of this book. Mr Fred Vance has contributed significantly to the artwork. Mrs A. Miller-Craig kindly checked the German references in the bibliography. As to funding, special mention must be made with gratitude to the James and Grace Anderson Trust for support of work in the field of cerebral palsy.

1
CONCEPTS, CLINICAL METHODS AND CONTRACTIONS

Inevitably we try to describe the intangible in terms of the tangible. Some everyday words describing mental states are based on mechanical analogies: *e.g.* agitated, antagonistic, apprehensive, cracked, depressed, driven, emotional, energetic, persecuted, pressured, repressed and strained. 'Trepidation' comes from the Latin, *trepidare*, 'to hurry' or 'to bustle'.

Muscle tone, too, is obvious in words such as attentive, rigid, slack, stressed, stiff-necked, tense, tensed and up-tight, while one of the Greek meanings of tone, associated with music, comes through in expressions such as 'keyed up' and, presumably, 'highly strung'. We are used to assessing mood according to deportment:

> Edward Heath, long hunched in self-imposed misery under the reign of Margaret Thatcher, is suddenly his old jolly self. His shoulders, now relaxed, are heaving in merriment. You know the sort of thing. 'Ho, ho, ho,' he is chuckling. Life is good once more. (*Sunday Express*, 26.5.91).

Charles Darwin (1872) made a detailed study of the physical manifestations of emotions (Fig. 1.1). Chronic anxiety is often reflected in overactivity of the general body musculature. Using the electromyogram (EMG), there have been investigations of the electrical activity of muscles in anxiety; although (to paraphrase a comment by George Bernard Shaw about Pavlov's experiments on conditioned reflexes in dogs) some of these reports say no more than what every policeman knows. There is increased electrical activity in the frontalis muscle in chronically anxious people, but after all the significance of a furrowed brow is lost on no one. Howell (1905) noted that:

> An increase in mental activity, so-called mental concentration, whether of an emotional or an intellectual kind, leads, by its effect on the spinal motor centers, to a state of greater muscular tonus, the increased muscular tension being, as it were, visible to our eyes.

Later in this book some consideration will be given to the relationship of mental state to muscle tone (pp. 56, 121).

The asymmetry of facial palsy is partly due to the muscles on the good side pulling the side which has no innervation. Owing to paralysis of the orbicularis oculi muscle the palpebral fissure is wider on the sound side and the lids may not close when the patient sleeps. Muscular tonus needs to be considered not only in large but also in small and inaccessible muscles. Occasionally a condition is found in which the eyelids droop. Sometimes the levator palpebrae superioris, which is in part composed of smooth rather than striated muscle, is weak from myasthenia

Fig. 1.1 *(left).* Portrayal of 'terror'. A woodcut from Darwin's famous book on the *Expression of the Emotions.* Some of the illustrations were of actors posing with what they took to be appropriate manners. This picture was copied from a photograph of a man whose muscles were being stimulated electrically; the experiment was performed by the French neurologist Duchenne (1806–1875), who made extensive use of electric shocks to demonstrate muscular actions. His name is associated with the hypertrophic form of muscular dystrophy (from Darwin 1872).

Fig. 1.2 *(right).* Bilateral ptosis associated with tabes dorsalis. By tilting the head backwards the person may be able to see through the narrow apertures that are still open (from Gowers 1886).

gravis, where there is a failure in the transmission from nerve to muscle or some other cause; there is ptosis (Fig. 1.2).

Muscular development

Everyone is accustomed too to estimating physique by appearance. We observe whether or not someone has heavy jowls, due to underlying powerful muscles and strong bones. In one dental department it was found that people with long faces could exert significantly less force during biting than people with normal physiognomy (Proffit *et al.* 1983). Everyone must know Longfellow's poem *The Village Blacksmith*:

> Under the spreading chestnut tree
> The village smithy stands;
> The smith, a mighty man is he,
> With large and sinewy hands;
> And the muscles of his brawny[1] arms
> Are strong as iron bands.

[1] Brawn is another name for muscle; the word originally meant a part of an animal, such as the calf of a wild boar, suitable for a roast.

Our ancestors relied heavily on their own muscle power and there is evidence from their bones that their muscular development was greater than is common today. Rubin (1974) wrote about skeletal adaptation:

> The fibulae show evidence to suggest that the muscles of the legs were robust and developed. In a group of Anglo-Saxons from Guildown, Surrey, for example, a roughened area was often found on the posterior aspect of the upper third of the fibula at the origin of the soleus muscle, indicating an active and powerful muscle.
>
> The bones of the foot similarly exhibit general evidence of the great strength and strong use of the leg muscles. The areas of muscle attachment on each bone are frequently more developed than is now the case and are the result of the pull due to strong muscular activity, as well as from the increased rapidity of limb movement required in former times. The inherent instincts of speed, alertness and rapidity of movement, not only in defence, but in hunting and other essential activities, could well have been vital for survival.

John Hunter (1728–1793) was a famous anatomist and surgeon who studied the development of teeth, the function of the air sacs in birds, the olfactory nerves, sex behaviour in animals, comparative aspects of the placenta and testis, the behaviour of bees and temperature variations in plants and animals. Writing on muscular development (Hunter 1837), he observed that:

> The habit of acting in a muscle, especially when employed in considerable exertions, increases the necessity of becoming stronger, which necessity, acting as a stimulus upon the muscles, becomes a real cause of increase of size, which augments their strength. This effort is so evident, that painters and sculptors, as well as physiologists have observed it. We have Charon[2] and Vulcan[3] always represented with large shoulders, brawny arms, and their lower extremities small, and apparently disproportioned. This effect is still more nicely marked by the difference between the right arm and the left; the right being generally employed in preference, and more particularly in great exertions, is therefore the largest and strongest. People who play much at tennis, where the ball is always struck with the right hand, have that arm much thicker and stronger than the left; therefore a man originally well-proportioned will lose that proportion by being employed in any action that does not require the whole body . . . The increase in the voluntary muscles appears not to be without limitation, and indeed if it was we might see them increase beyond conception.

A casual glance at a magazine devoted to 'pumping iron' shows that the limitation of muscular development by exercise is not trivial! Charles Atlas made a fortune by selling descriptions of a series of exercises which could be performed without special apparatus, no weights, no springs. A slogan such as 'you can have a body like mine' was placed alongside a photograph of his bronzed and muscular figure together with statements such as 'I was a 97 pound weakling' (Butler and Gaines 1982).

Muscle development has been intensively studied. It seems that to increase strength, high tension is needed during the training programme. It may not be very important just how this is achieved, and it need not be fatiguing. There is evidence that a mere 10 contractions daily at loads which are nearly maximal can, if repeated regularly, really increase the strength of the muscles involved.

---

[2]Charon—a figure of the underworld who rowed the dead across the river Styx.
[3]Vulcan—god of fire of the ancients who was lame but reputedly highly skilled in the use of hammer, anvils and other ways of working metal.

Fig. 1.3. Generalised weakness associated with tabes dorsalis. There was subluxation of some of the joints of the body, the ligaments no longer holding the bones in place (from Charcot 1881).

Fig. 1.4. The sequence of movements in 'Gowers' sign' (from Gowers 1886).

The greater muscular strength of men is naturally attributable to the male sex hormone. There seems to be no doubt that compounds which simulate its actions, anabolic steroids, are of use in restoring muscle size and strength in people who are emaciated. However, misguided athletes have often taken anabolic steroids in the hope of building up lean muscle mass and improving performance. Brown and Benner (1984) reviewed the evidence and found a number of conflicting reports:

> Generally, studies which did not support athletic steroid use were well designed, whereas those that did tended to have serious flaws in study design, analysis, or calculation . . . A review of the literature evaluating the effects of anabolic steroids on athletic performance suggests that benefits are minimal or non-existent. Steroids, administered under the best of circumstances, using 'safe' doses, in conjunction with an exercise training programme and a high-protein, high-calorie diet, have not resulted in increased athletic ability.

Rather similar conclusions were reached by George (1988). He pointed out that weight gain and increased muscle girth might be due merely to the retention of water and salt. The taking of these drugs may have unpleasant and indeed dangerous side-effects, and his review of the possible consequences makes for sombre reading.

The amount of muscle in a limb can be estimated by computed tomography (Maughan *et al.* 1984). In a study of young adults it was found that the males had on average 72 per cent muscle, 15 per cent fat and 13 per cent bone in the forearm. The corresponding values for the females were 59 per cent muscle, 29 per cent fat and 12 per cent bone. The differences in muscle and fat composition were highly significant. It is possible to pick up a fold of skin and measure its thickness by calipers. Then one can estimate the percentage of forearm fat content as follows, using the data for a skin fold in millimetres over the biceps:

Males: % fat = skin fold × 2.8 + 2.7   (equation 1)
Females: % fat = skin fold × 3.3 + 7.5   (equation 2)

Having obtained these values it is possible to estimate the muscular content of the forearm, M, if the volume, V, between the wrist and elbow is measured by water displacement:

Males: $M = V \times 0.92 - \text{fat volume} \times 0.97 - 38$ (equation 3)
Females: $M = V \times 0.49 - \text{fat volume} \times 0.61 + 205$ (equation 4)

The values used in these four equations are in millimetres. These relationships are merely empirical and the numbers on which they were based were small, but they may be of some use.

Under some circumstances the muscles become very weak (Fig. 1.3). Loss of power may result from damage to the nerves as in polyneuritis. With weakness of the legs due to muscular dystrophy, the patient may have to 'climb up' with his or her arms (Fig. 1.4).

Etymology
Voluntarily controlled muscle makes up about 25 per cent of the bodyweight at

birth and about 45 per cent in the adult man, and the condition of the musculature has powerful implications for good or ill. Inevitably much of this book is about muscle tone. The term 'tone' has a very long history, and one aim of this work is to give a modern interpretation of phenomena broadly covered when physicians and physiologists refer to it. A good understanding of the subject has wide practical implications. A physical medicine specialist (Clemmesen 1951) wrote:

> The concept . . . has often been discussed in relation to physical education, remedial exercises, posture correction, economy of work, and the mechanics of locomotion. In rheumatology it influences our conceptions of contractures, reflex contractions of the muscles, muscle spasms, and the tender palpable nodules of the muscles, the so called myalgic spots. The different modern procedures called relaxation have to be evaluated in relation to our knowledge of physiology. We must be able to analyse this concept before we can select suitable physiotherapy for an individual case.

Knowledge of biomechanical principles has been meagre until relatively recently, and this comparative ignorance has been responsibe for many of the ambiguities surrounding the use of the term. Fenn and Garvey (1934) succinctly posed basic questions on this topic:

> Muscle 'tonus' has been described, defined, and measured in a multitude of different ways. Scientifically speaking, however, there is no such single property of a muscle as its tonus. Rather tonus is a convenient term which includes many different properties such as elasticity, viscosity, and muscle reflexes. The continued employment of the term, convenient as it is, serves to avoid the necessity of analysing 'tonus' into its various components. An experimental analysis of this sort however is exactly what is needed for a complete understanding of the phenomenon and for an intelligent and scientific classification of the various types of rigidities familiar clinically.

The origin of the word 'tone' can be traced in Liddell and Scott's *Greek-English Lexicon*, and I am indebted to E. K. Borthwick, Emeritus Professor of Classics at Edinburgh University, for the following account:

> The verb 'I stretch', τείνω (teinō), is the common form and is used by Homer of stretching of a bow, reins, *etc.* also 'stretch oneself' in running. Aeschylus uses it of straining the voice. Galen uses it of stretching tendons *etc.*
>
> The noun, τόνος (tonos), is apparently first attested in Xenophanes (sixth century BC philosophic poet) of exertion or striving after 'virtue'/'courage'. It is used by Aeschylus of stretching flax; in Herodotus and Aristophanes of bed and chair cords, in Plato and Aeschines of pitch of voice, or accent; in Aristoxenus and subsequent musical writers of pitch-key; in the medical writer Soranus (second century AD) of power of contracting muscles.
>
> The adjective, σύντονος (suntonos), 'stretched tight', is common in music (of lyre strings, high pitch, *etc.*) and of muscular strains, *etc.*

The idea that the word is connected with tuning a musical instrument foreshadows findings (described later in this book) of the relationship between tone and the resonant frequency of the limb; indeed the word 'tune' has the same root as tone.

The origin of the word 'spastic' is also Greek. Homer used the basic verb σπάω in the sense of drawing a sword. The idea of a muscle pulling was there from the start. In Herodotus, Euripides and Plutarch it referred to causing a spasm. Σπασμωδες was used to describe attacks of cramp. The adjective σπαστός is not cited at all except with the prefixes ἀνα, ἀντι, δια and ἐπι, but there is no reason to

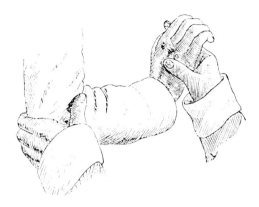

Fig. 1.5. The type of manipulation customarily used to test for tone. The illustration originally was made for a different reason. With lead poisoning the long extensors of the forearm may be paralysed but if the proximal phalanges are extended by the examiner the patient can powerfully extend the two distal phalanges by the action of the interossei muscles (from Duchenne 1867).

suppose it was not used by itself. The root of 'rigid' is Latin *rigidus, -a -um*, meaning 'stiff', 'hard', 'inflexible', 'rocky'. 'Muscle' comes from the Latin *musculus*. This is a diminutive of *mus*, 'a mouse'. Contract your biceps and you can perhaps imagine a small mouse running along under the skin.

Tonics
The use of 'tonics', substances expected in some undefined way to improve performance, may now be regarded as outdated. Some preparations contained small doses of strychnine. Most of the iron tonics sold over the chemist's counter were bought because iron was equated with strength, rather than because of a popular awareness of the needs of haemoglobin. At least in anaemia they would have been useful; whilst glycerophosphates, at one time very popular, probably cure no ills. There are dissenting voices, however, claiming usefulness for other preparations of tonics (Fulder 1982).

Clinical methods of assessing muscle tone
There have been multitudinous descriptions of the resistance felt on moving a limb (Fig. 1.5). Foster (1890) gave the following account:

> When we handle the limb of a healthy man, we find that it offers a certain amount of resistance to passive movements. This resistance . . . is an expression of muscle tone, of the effort of the various muscles to maintain their 'natural' length. In many cases of disease this resistance is felt to be obviously less than normal; the limb is spoken of as 'limp' or 'flabby'; or as having 'a want of tone'. In other cases of disease, on the other hand, this resistance is markedly increased; the limb is felt to be stiff or rigid . . .

Most physicians and surgeons treating patients with disturbances of muscle tone will not have any special instrumental methods available, so it is worth describing some clinically useful procedures.

TABLE 1.I
The Ashworth scale

Score	Degree of muscle tone
1	No increase of tone
2	Slight increase of tone, giving a 'catch' when the limb was moved in flexion and extension
3	More marked increase in tone, but limb easily flexed
4	Considerable increase in tone—passive movement difficult
5	Limb rigid in flexion or extension

From Ashworth (1964).

The Ashworth scale
Dr Bryan Ashworth, of the Department of Clinical Neurosciences of Edinburgh, devised a scale which can rapidly be used and which is widely employed (Table 1.I).

The Wartenberg test
Wartenberg (1951) pointed out that the physician in his office needs to resort to the 'old rather crude' method of moving the limbs whilst attempting to appraise the resistance he feels. Simultaneous examination of the two sides is impossible, comparison is difficult and minimal changes can hardly be detected. In an excellent clinical account he described the test which has come to bear his name:

> The patient sits on the edge of a table with his legs hanging freely. The examiner lifts the patient's legs simultaneously to the same horizontal level, then releases them, permitting them to swing freely . . . There is a characteristic precision and regularity in the motion of normal free-swinging limbs. The range of successive movements diminishes steadily and evenly.

The swings are reduced, often greatly so, in Parkinsonism, and an asymmetry can be a useful early diagnostic sign in unilateral Parkinsonism.

> In pyramidal spasticity, movements are jerky and irregular, the forward movements are brisker than the backward, and the range of backward movement is diminished. The most outstanding characteristic, however, is that the limb does not swing in a precise anteroposterior plane. Movements are zigzag; the tip of the foot describes irregular patterns such as a flattened ellipse, a broken circle, a spiral or an indefinable figure. It is sometimes possible to detect spastic hemiparesis on the basis of this test alone.
> Hypotonia of any origin influences the quantity of movement; the overall swinging time of the affected leg is longer. This occurs in affections of the cerebellum and of its connections. This increase occurs also in any involvement of the lower motor neurone.

Since this description, there have been many investigations of these phenomena using instrumental assistance. These studies are described later (pp. 30–33) but all the accounts I have come across describe monitoring of only one side; Wartenberg's emphasis on the importance of simultaneous comparison of the two sides has been disregarded.

The work of André-Thomas
Two French neurologists, André-Thomas and de Ajuriaguerra (1949), published a

Fig. 1.6. Tests for extensibility of the upper limb. The stretching is undertaken quite slowly and the range of motion noted. (From André-Thomas and Ajuriaguerra 1949, by permission.)

large work of classic importance dealing with clinical variations of muscle tone in a wide variety of clinical conditions. There is also a short account of the work in relation to the neurological examination of the infant which is in English (André-Thomas *et al.* 1960). There are three fundamental types of procedure.

1. MUSCLE CONSISTENCY

This is investigated in two ways. First, the muscle is palpated and a difference is likely to be felt between the healthy and the affected sides. Second, the limb is shaken and the transverse displacement of the muscle is observed. The muscle shakes like a piece of cloth when the firmness is diminished, and moves little when it is increased.

Instrumental attempts to investigate the consistency of muscles are described in Chapter 2 on pages 23 and 24.

2. *EXTENSIBILITÉ*

This is the capacity of a muscle to be lengthened. It is revealed by the range or amplitude of slow passive movement of a segment of a limb at a joint. The procedures adopted for testing different muscle groups are shown in Figure 1.6.

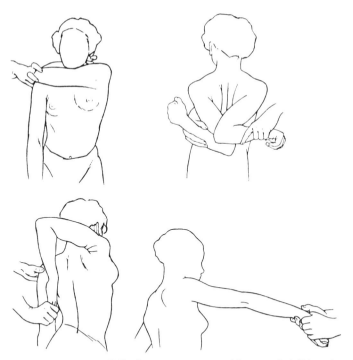

Fig. 1.7. Hyperextensibility in a young woman with congenital dislocation of the hip. It is not uniform. The triceps is not sufficiently extensible to allow the wrist to be put in contact with the shoulder; the flexors of the elbow are however lax as are to a striking degree certain of the other muscle groups. (From André-Thomas and Ajuriaguerra 1949, by permission.)

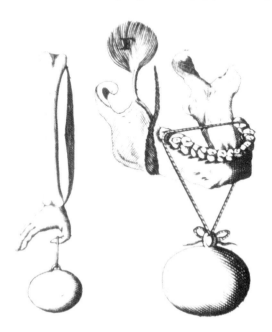

Fig. 1.8. Thumb and jaw carrying loads, a study of static considerations (from Borelli 1685).

Occasionally, excessive range may be a result of pathological connective tissue changes. Findings in such a patient are shown in Figure 1.7. A further discussion of the range of possible movements will be found on pages 103–108.

Bird and Stowe (1982) studied the range of motion possible at the wrist joint in different directions, using a simple protractor-like instrument, a 'goniometer'. They found that the range of motion was greater in females than in males, and was always greater for passive than active movements.

3. *PASSIVITÉ*

The French workers took particular note of the resistance to movement imparted to distal segment by 'flapping' and observed the extent of the resultant motion. Asymmetry was assessed. This shaking procedure is naturally a rapid stretching of the musculature; its instrumental counterpart, the effect of applying rhythmic forces to the limb as a method of assessing tone, is described later. Rapid stretching is effective in activating the phasic stretch reflexes which are often a prominent feature of spasticity.

Tonic postures

It is easier to make deductions about static than about dynamic conditions. Giovanni Alfonso Borelli (1608–1679), the father of biomechanics, wrote a famous book in which the postural system was considered in terms of weights, levers, pulleys and the muscular tensions which inevitably have to be involved in holding postures. Such analyses were, and inevitably remain, fundamental (Fig. 1.8).

There are different ways of arranging the fibres in muscles. Unipennate muscles have a single set of oblique fibres. The extensor digitorum longus has a tendon running along one side. In the 'bipennate' arrangement, in the rectus femoris, the tendon passes up the centre of the muscle and the fibres are attached obliquely on either side. Others are 'multipennate'. In the tibialis anterior there are aponeuroses of tendinous material approaching the belly of the muscle from both ends and the fibres run only a short distance.

The word pennate is derived from the Latin *penna* ('a feather'); the resemblance is obvious. In pennate muscles more fibres are packed in a shorter distance than muscles with straight arrangements. They may thus be comparatively powerful although the gain is offset by the pull being at an angle. The range of possible movement is reduced.

The variations in the medical use of 'tone' since the time of Galen of Pergamus (c.129–199 AD), the Greek physician who founded experimental physiology, were described by Sherrington (1919). He wrote:

> It is clear that for Borelli the term *'tonic'* still conveyed *'postural'*. In short, the word had lived without change for nearly sixteen centuries, and even a century of the Renaissance had not surcharged it with more, or at least with altered meaning.

Often, then, the term has been applied to the tension produced by an active

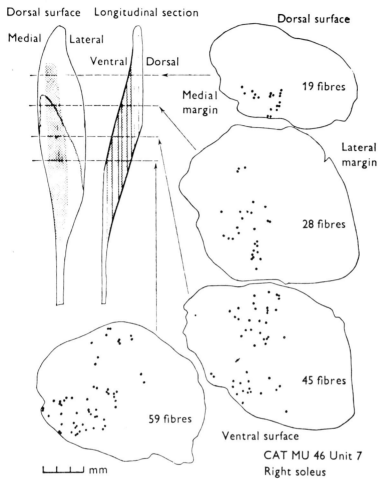

Fig. 1.9. Representation of distribution of muscle fibres innervated by a single nerve fibre. The levels of sectioning are indicated on the whole muscle diagrams. (From Burke *et al.* 1974, by permission.)

muscle in holding a particular posture as when the musculature has to resist gravity. Bernstein (1967) noted that:

> Invertebrate organisms have in their make-up a form of co-ordinational surrogate in mechanisms of muscular locking (Sperrung) which by physiological means eliminate such degrees of freedom as are unnecessary at any given moment. But we must add that all lower forms of vertebrates (up to birds inclusively), for which the striatum still predominates over the cortical hemispheres, possess analogous auxillary muscular mechanisms and employ them widely. Lizards, snakes, many brooding birds (eagles, parrots, *etc.*) are as rigid as statues in the intervals between voluntary movements. Reptiles show particularly clearly a statue-like stiffening of the body as soon as successive voluntary movements cease. If a lizard turns head to tail its body and limbs are motionless as sculpture . . . In the norm there is no rest in mammals and in human beings, and outside of deep sleep there is no similar immobility; careful observation of standing or sitting human beings, dogs or cats gives evidence of this. Even the set immobility of a cat or a tiger is quite unlike the immobile period in a reptile (or spider)—it is sufficient to watch its tail.

The concept of active tonus covers in principle much of motor control which it is not feasible to review comprehensively in a book concerned mainly with resting tonus. Just a few aspects have been selected. Some aspects of the control of voluntary movements are considered in Chapter 11.

Motor units

There are some 434 muscles in the human body and these contain about 250 million individual striated muscle fibres. In the muscle of the eye they have a diameter of about 20μm; in the muscles of the limbs they range in size from 10 to 100μm, being largest in athletic men.

A number of muscle fibres are innervated by a single nerve axon. The number controlled in this way varies, being least in muscles such as those of the hand concerned with fine movements, and greatest in the large muscles of the trunk. The distribution of muscle fibres innervated by a single nerve fibre has been studied in the cat. Single motor neurons have been stimulated electrically and the resulting depletion of glycogen acts as a marker as to the muscle fibres innervated. Muscle fibres belonging to single soleus motor units were found to be scattered through territorial volumes occupying a large fraction of the total muscle volume (Fig. 1.9). Between 50 and 400 muscle fibres were found to be innervated by a single nerve fibre (Burke *et al.* 1974).

Different types of muscle fibre

There are different types of motor unit and everyone knows that some muscles are pale and some are red. Pale muscles act rapidly, while red muscles are used for holding postures and continuous low-level action. If a grilled trout is laid on its side and the skin removed, a thin line of red muscle is visible. It is this alone which is used in normal cruising. The large bulk of the fish is composed of the pale fibres which are needed only when speed has to be increased.

The red colour is due to the pigment myoglobin which acts as a short-term oxygen store so that contraction can be maintained if the muscle, by its own contraction, shuts off its blood supply. The store lasts for just a few seconds. There have been surveys of the composition of muscles in different animals. A greyhound is said to have only about 3 per cent of red fibres; a mongrel may have 30 per cent. Pale fibres are of two types; one rapidly fatigues, the other does not.

The most rapidly performing of any of the muscles in the body are those that move the eyes. The slowest contractions of any are those of the soleus. As the weight line of the standing human body normally falls in front of the ankle joint, the soleus is often active for some seconds at a time to prevent the torso falling forwards. Most people have about 75 per cent of slow fibres in their soleus muscle, though all are slow in some cases. Sale *et al.* (1982) measured the timing of twitches of the muscle after electrical stimuli. The 'contraction time', from the start of contraction to the peak tension was found to be 112 ± 11ms (mean and SD). Relaxation was slower. The 'half relaxation time' was measured, *i.e.* the time needed for the tension to fall from the peak to 50 per cent of that value. It was 110 ± 14ms.

The often numerous fibres supplied by a single motor nerve fibre are all of the same type. The composition of a muscle can be determined by histochemical techniques which show up the presence of a variety of enzymes.

Sounds generated by contracting muscles

If a stethoscope is applied over a muscle which is then contracted, a rumble is heard; there are generally many motor units that are active. If the index finger is placed beside the eye, and the eye is partially closed, a vibration is felt. Muscle activity is of course discontinuous; in the limbs the discontinuities are averaged out because many units are firing asynchronously and also because of the considerable effects of the inertia of the limb segments. Gordon and Holbourn (1948) were interested in these effects:

> The most satisfactory form of 'stethoscope' is an 8in. length of 3mm. bore pressure tubing, one end of which is inserted into the ear and the other, cut accurately flat and preferably wetted, placed flat on the upper eyelid. With this instrument, the sounds from single units are heard with such clarity as to justify the use of this experiment in class teaching. The sounds resemble those of a distant motor-cycle.

Gordon and Holbourn were also interested in the sounds of muscular origin which a person can hear in their own ears without special apparatus:

> One of the simplest ways of hearing the muscle sounds is to close the external meatus at its outer end, for example by lying with one's head on a pillow. Once one has learnt to recognize these sounds, they are heard in this way with great clarity . . . The sounds may also be heard well when the outer end of the meatus is blocked with soap suds, or when the meatus is filled with water. It is frequently possible to hear sounds like the rhythm from a single motor unit. They occur when the jaw is closed or deviated to the same side as the ear one is using, and cease when the jaw is opened or deviated to the opposite side: they probably arise in m. masseter. The lowest rate of firing is about 5 per sec., and occurs during slight maintained contraction. It sometimes seems impossible to achieve complete relaxation even with one side of the head lying on a pillow. At rates below about 5 per sec. the rhythm becomes irregular.

Some people can at will induce a muscular type of noise in their ears and render themselves partially deaf for a short time. When I used to lecture on hearing, almost every year students would tell me that they had developed this ability in childhood. One had used it at school to muffle sharp words from the teacher. Some thought that everyone could perform the manoeuvre. Two reports are characteristic.

One was a 22-year-old white American who could click his ears momentarily or induce a sustained effect for up to 15s. He could do this when no other muscular action was apparent except for a suggestion of a movement of the hyoid. The noise to him was similar to that of an EMG heard through a loudspeaker. He noted that the movement reduced the discomfort of loud noises. If he saw that a door was about to bang he induced a contraction. If he entered a cinema towards the end of a film he could reduce the clarity of the speech to an extent that it could be ignored.

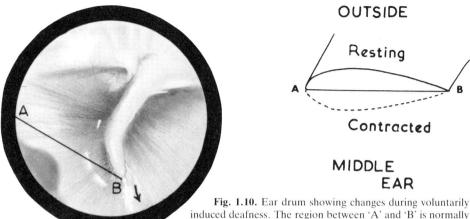

Fig. 1.10. Ear drum showing changes during voluntarily induced deafness. The region between 'A' and 'B' is normally convex outwards; during the manoeuvre it is sucked in and becomes concave outwards. (Arrow indicating light moves down.)

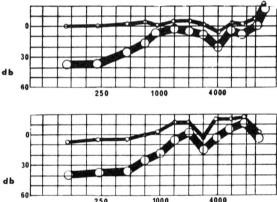

Fig. 1.11. Audiograms, 20-year-old male. *Top:* right ear, and *bottom:* left ear. The heavy lines represent the sensitivity during the manoeuvre.

Another was a 23-year-old half-Chinese student. At the age of eight, while living in Japanese-occupied Borneo, he was in an area subjected to bombing by American planes. During the bombing he found that he could, by concentrating, produce a sound like that of distant thunder and reduce the discomfort of the explosive noises. On auscultation a muscular sound could be heard arising from the ears.

In students who could manage this trick, otoscopy revealed a reduction in the reflection of light from Shrapnell's membrane. It is difficult to see motion away from the otoscope without some aid. Accordingly talcum powder was introduced, so that a few white particles adhered to the drum; retraction then became plainly visible (Fig. 1.10). A loud sound introduced into the opposite ear using a Bárány box did not induce the changes. As there was no motion of the malleus, it seems likely that the muscle that contracted was not the tensor tympani but the tensor palati. Audiometry showed that there was a loss of hearing for the lower frequencies (Fig. 1.11), while at higher frequencies the principal alteration was a change in the apparent pitch of the note[4]. There was no change for bone conduction. Measurements were made using a simple water-filled manometer of the change of volume corresponding to the alteration of the shape of the drum.

These circumstances, where an aspect of motor performance is established late enough for the person to describe the events, may perhaps give a clue to the ways in which the wide variety of controls of the musculature are established between birth and the age of five.

Active tone in large muscles

Postural tone is a subject which has long interested physicians and other therapists. Delpech's method of improving posture involved sitting with a band round one's head (see Keith 1919). The beak of a swan mounted on a pillar carried a pulley and connected the band on the head with a weight. The muscles were exercised by having to act against the resultant resistance. The tension was said to be sufficient to keep the muscles which maintained the spine 'on the alert'!

A muscle which contracts is liable to cut off its own blood supply, at least if a certain tension is reached. Muscle fatigue is relatively uncommon in everyday living. The reasons have been discussed by Ralston and Libet (1953):

> Probably the main reason that fatigue does not occur under ordinary conditions of rest or moderate activity is simply that our muscles are not continually active. Not only is there no persistent background activity in normal resting muscles generally, but even in those muscles that may be actively involved in maintaining some postural attitude other than a relaxed one, the periods of activity appear to be generally of minimal duration. If one will watch the behaviour of his own body and that of his neighbour throughout the day, one will see that no one postural attitude is or can be maintained for any appreciable period of time. After periods measured in seconds, we alter our posture so as to avoid the necessity of using any one muscle for more than a short time. The head, for example, is rested on one hand, and then on the other, or dropped towards the chest. In the latter case, it is supported purely by the elastic forces of the muscles and other tissues of the posterior region of the neck.
>
> Furthermore, even in an active motor pattern such as locomotion, involving displacements of the body and its parts, it has been found . . . that any given muscle is active only for a brief period, of the order of a fraction of a second generally. It is the proper sequence of many muscles each acting in a brief burst that gives the smooth over-all pattern. Thus one can walk for long periods without fatiguing any particular muscle.

Certainly the early photographers were aware that their subjects needed to be restrained when an exposure lasting some minutes was necessary (Fig. 1.12). This

---

[4]It has long been known that if the pressure in the middle ear is either increased or decreased there is impairment of the ability to hear low-pitched sounds. 'The effect of forcing air into the tympanum is exactly the same as that of exhausting it, with reference to the perception of grave and acute sounds. Thus the dull rumbling sound of carriages passing over a bridge, or the firing of cannon, cease to be heard on the tension of the membrani being induced in either of the ways mentioned. But the treading of horses upon the stone pavement, the more shrill creaking of carriages, and the rattling of paper, I hear very distinctly while my membrana tympani remains tense. The character of the deafness is very strikingly shown by the fact that the ticking of a watch is heard at the distance of eight feet quite as distinctly when the membrana tympani is rendered tense as in the natural state, perhaps even more distinctly, while all dull grave noises in the street cease to be heard' (Müller 1842).

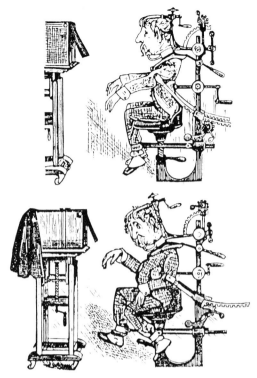

Fig. 1.12. Early photographic portraiture required long exposures. The clamps used for keeping the person still were the subject of these cartoons.

account, dated 1841 and quoted by Stevenson and Lawson (1986), was not originally intended to be funny:

> Paint in dead white the face of the patient, powder his hair and fix the back of his head between two or three planks, solidly attached to the back of an armchair; wind up with screws.

When the drugs known as phenothiazines (*e.g.* chlorpromazine) are used to treat mental problems, a condition of motor restlessness sometimes develops. There is a feeling of muscular quivering, an urge to move about constantly, and an inability to sit still. This is called 'akathisia' from the Greek a ('not') and καθίζειν ('to sit').

The vestibular apparatus

Many factors influence active tone. One example of a system which has powerful effects is the vestibular apparatus. The galvanic stimulation of this system makes a good student demonstration.

Electrodes, prepared by covering metal plates of about 2cm diameter with cotton wool and gauze, are soaked in saline. The subject stands with eyes shut and feet together and the electrodes are applied to the mastoid processes, one on each side. When a current of 1mA is passed, the person falls to one side and may need to be caught. If the direction of the current is reversed, s/he falls the other way. Provided that the electrodes are held quite firmly on the skin there need be no

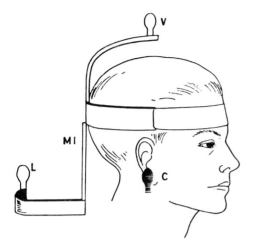

Fig. 1.13. The person wears a helmet carrying two lamps, one at the vertex (V), one behind the occiput (L). Mirror (M1) forms an image of lamp L at the base of the skull. Thus as the head moves on the upper vertebrae the vertex and occipital lamps are displaced but the virtual image remains steady. Beams reach mirrors. From the deflections recorded on moving the photographic paper it is possible to deduce the movements of the head (from Walsh 1963).

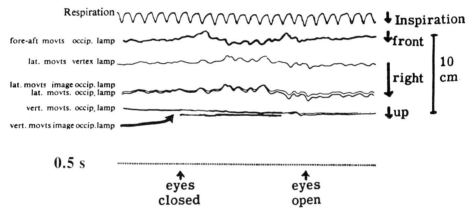

Fig. 1.14. Postural sway, in a 21-year-old female medical student with bilateral vestibular failure. The small slow variations of head position during standing are not unusually large (from Walsh 1963).

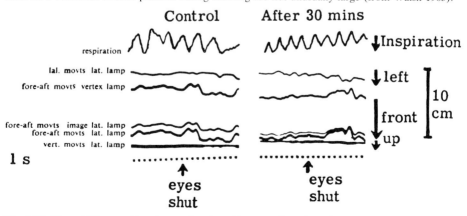

Fig. 1.15. Normal 20-year-old male standing in cold water; there is no increase of sway (from Walsh 1963).

discomfort or local sensation from the passage of the current. The excitation of vestibular receptors, probably principally the otolith organs, has immediately caused a striking redistribution of postural tonus.

With more vigorous stimulation of the vestibular apparatus, one arm may be thrown upwards, like an athlete throwing a discus.

Postural sway
The living body is never perfectly still. There is motion related to the movements of the chest and diaphragm in respiration, to the heart beat, minor muscular adjustments, visceral activity and so forth. If a sensitive recorder is fixed to any part of the body, microvibrations are always recorded. A person's frame is only immobile with the stillness of death. Many people have recorded the normally small movements which are continously evident when someone stands. Such postural sway can be recorded by getting the person to stand on a platform supported by stiff metal springs. These are flat and the distortion is registered electrically by strain-gauges; a variety of 'ataximeters' are commercially available. The author recorded postural sway during standing using an optical apparatus (Fig. 1.13). A different version has been used to record head movements in trains (see Fig. 11.1).

Even with loss of the vestibular apparatus a person can still stand in an approximately normal manner (Fig. 1.14). The particular subject tested was however very susceptible to small tilts and, walking in her own home at night, was liable to lose her balance if she stepped on a loose floorboard. Cooling the ankles to reduce joint sensibility does not increase sway (Fig. 1.15). Position sense is now believed to be principally a function of receptors in muscles (p. 103).

Paulus *et al.* (1987) studied a patient with loss of position sense, due to diabetic polyneuritis, and vestibular failure following treatment with an antibiotic, (gentamicin) for osteomyelitis. With the eyes open he could stand and walk slowly. With the eyes closed he overbalanced within one second. In normal subjects postural, vestibular and visual signals normally provide redundant information about posture. Reduced to vision alone this patient intuitively searched for nearby visual objects for reference.

Tone in relation to skeletal deformities
The founder of orthopaedics was Nicolas Andry (1658–1747). He coined the word orthopaedics from the Greek ὀϱθο and παιδεία—literally 'the straightening of children'. His book on the subject was published when he was in his 80s (Andry 1743). He grasped the important truth namely that muscles were the chief instruments in shaping a child's body; it was therefore by playing on these instruments that a physician could accomplish his orthopaedic aim.

Section of tendons for the treatment of deformities was introduced by Jacques Delpech (born 1777), whose studies of people with club foot were described by Keith (1919):

Fig. 1.16. 'Young girls ought not to be allowed to sew or read, except in an erect posture; they should hold their work or their book to their eyes, and not their eyes to their work, without which their body will infallibly become crooked' (from Andry 1743).

One was a soldier in whom there was paralysis of the muscles which dorsiflex and evert the foot, owing to a wound of the external popliteal nerve. In the course of time Delpech witnessed the soldier's foot gradually assume the characteristic club form. Delpech realised then that club-foot must be the result of the unopposed action on the part of the sound muscles of the leg—the muscles which plantar-flex and invert the foot.

The Alexander method

Active tonus is of interest to lay therapists, and it is difficult to describe the methods or rationale, for their practical manoeuvres are laced with *ex cathedra* statements. The Alexander method concerns the 'correct' ways of moving, sitting, standing and so forth. Anyone interested should read the accounts of the technique themselves, or preferably search out an 'Alexander teacher'. Hodgkinson (1988) wrote:

> Although the Alexander Technique is not primarily a self-help therapy, it is possible to bring the basic principles into everyday use, without having lessons.
> The three most important Alexander principles establish primary control. They are:
> - Let the neck be free. Never increase the muscle tension in the neck.
> - Let the head go forward and up, never back and down to sit on and crush the spine.
> - Let the torso lengthen and widen out. Do not shorten the back by arching the spine.

To its many devotees the system evidently induces a feeling of well-being. Certainly some postures are more becoming than others (Fig. 1.16).

Conclusions

Appropriate scientific investigations alone can provide machine-type data and reveal a wealth of knowledge about the mechanisms of muscle action and tension of which the unaided observer is inevitably unaware. As far as muscle tone is concerned, an experienced human examiner can very rapidly form assessments of the resistance to motion at different joints. Comparisons, however—often needed for assessing the efficacy of the treatment of spasticity and rigidity—are difficult, for the consistency of the observations over a long period of time is questionable. Most people, including many physicians and physiologists, would scarcely consider that the underlying processes involved when a limb is moved passively would deserve more than momentary attention. Later in this book it will be shown that this view is erroneous.

2
HISTORICAL AND ELEMENTARY METHODS OF MEASURING MUSCLE TONE

The skeletal muscles have been described as being placed 'on the stretch' in the living body. If a muscle is cut away from its attachments at each end, it may shorten; if a tendon is cut across, the ends often separate. Naturally, if the muscle contracts momentarily, the tendon will be drawn away from the wound and then there is no restoring force to drag it back. How far the retraction of tendons is due to this factor and how far to inherent elasticity can only become clear if the features of the anatomical arrangements of the muscle and the position of the limbs are considered.

Each fibre is made up of a chain of sarcomeres[1]. If overstretched to more than about 3.5μm, or if shortened to about 1.25μm, the sarcomere can generate no force. The muscles in contraction may be expected to shorten by around 60 per cent of their length fully stretched. In muscles of 'full action', this is quite adequate for the range of movement allowed for by the bones to which the muscle is attached. With muscles of 'short action', however, the fibre length is so short that they do not remain taut throughout the full range of movement permitted by their attachments.

With the ankle fully plantarflexed and the knee flexed too, the Achilles tendon is slack. The soleus and gastrocnemius are therefore examples of muscles of short action. The tibialis posterior and the peronei are the most important flexors of the ankle joint in that position. Other muscles of short action are the biceps femoris, semi-membranosus, rectus femoris, plantaris and the long head of biceps (Haines 1934).

The extent to which a muscle may shorten will be a function of the obliquity of the fibres, being less in those of the pennate variety (see p. 11). It will also depend on the extent to which the length is occupied by tendon (p. 166).

In other words, does the muscle in the living body possess a latent tendency to shorten, which is continually being counteracted by its disposition and attachments? The extent to which this prestretching exists has many implications for neurologists, orthopaedic surgeons and so forth.

Measurements of muscle tone have as one aim the obtaining of relevant information about this tension, and a variety of procedures have been adopted by different workers. One function of this monograph is to explore and to attempt to evaluate the merits of different systems. The introduction of the clinical thermometer in the 19th century enabled fevers to be adequately classified for the

---

[1]From the Greek σαρξ ('flesh') and μέρος ('part').

first time; it is no longer considered adequate to assess body temperature by feeling the forehead!

Measurements of intramuscular pressure

Since the time of William Harvey (1578–1657) it has been known that each muscle, as it contracts, squeezes some of the blood from its capillaries into the veins and onwards to the heart. Henderson (1938) measured the pressure in a muscle in the expectation that this would serve as an estimate of resting muscle tone. In healthy young men the pressure in the biceps was 600 to 900 N m^{-2} (60 to 90mm water). In patients several hours after major surgical operations the pressure was 300 to 600N m^{-2} (30 to 60mm water), but detailed figures were not given, nor were any statistics apparently calculated. In moribund animals the pressure decreased progressively to zero. A similar apparatus was used by Kerr and Scott (1936) who measured muscle pressure in a wide range of medical conditions, perhaps strangely not including neurological conditions.

Since this time somewhat more advanced instrumentation has been used to monitor intramuscular pressure. A comparison of certain techniques was undertaken by Styf and Korner (1986). The methods are reviewed in a monograph (Matsen 1980). There seem to have been no further investigations of pressures in relation to resting muscle tone.

In some people, particularly athletes, the swelling of muscle with exercise may cause a troublesome rise of pressure as the space is constricted by fascial planes. In the chronic compartment syndrome a rise of intramuscular pressure may cause pain and impair muscle performance. Styf *et al.* (1987) studied patients with this condition. Muscle blood flow was normal at the start of exercise but decreased later, and the pressure during relaxation pauses became elevated. The treatment is to perform a 'fasciotomy', cutting the constricting connective tissue. Raised intramuscular pressure may occur also with fractures, giving rise to the 'acute compartment syndrome'. Volkmann's ischaemic contracture of the muscles may result.

Hardness measurements based on squeezing or rebound

Consciously or unconsciously, everyone is used to estimating the hardness or softness of everyday objects by applying pressure and sensing the amount of give. A similar system has been used for muscles. Laarse and Oosterveld (1973) used a 'myosclerometer' and attempted to measure muscle tone by compression. Yet because the procedure was static, there could have been a change of stiffness during the measurement, and the effects of motion which could show up properties related to a dynamic system (such as viscosity) could not be seen. This method does not appear to have been used to assess spasticity and rigidity, and the approach is singularly limited.

A somewhat more interesting system makes use of a hammer which strikes the belly of the muscle and then rebounds. The greater the stiffness, the quicker the

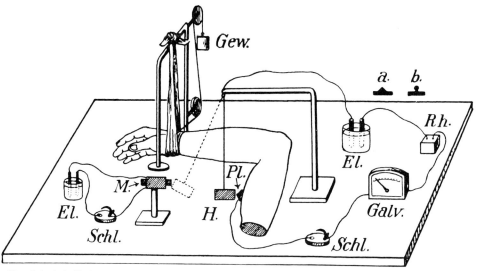

Fig. 2.1. A ballistic sclerometer. A hammer strikes a metal button fixed over the biceps. During the contact a current flows; the deflection of the ballistic galvanometer reflects the duration. The hammer is caught by an electromagnet. Interesting—but again unpromising! (From Springer 1914.)

rebound (Fig. 2.1). The method was refined by Simonson *et al.* (1949) who measured not only the time but also the height of the rebound. However, they did not succeed in obtaining absolute values for muscle constants; nor does the method appear to have been employed to investigate patients with spasticity and rigidity.

Semi-quantitative investigations using manual displacements
A number of investigators have sought to relate the tension in the musculature with displacement, moving the limb segment by hand, while obtaining the required relationship with some appropriate instrumentation. An early and sensible arrangement which could be used for measuring muscle tone at the elbow or knee was that of Schaltenbrand (1929):

> In order to avoid the influence of gravitation, the axis of the joint is put vertically in space. The extremity examined is hung by means of leather cuffs to two bars which are joined by a ball-bearing junction the axis of which corresponds, as far as possible, to the axis of the joint. The proximal bar is fixed in space, so that only the distal bar with the distal part of the limb can move. It is moved by means of a third bar, which rotates around the same axis but is connected with bar 2 only by two dynamometers[2], one for the extensor and one for the flexor side. The record of the dynamometer is written on a kymograph by means of a thread conducted through the axis of the instrument . . . The position of the moving part is recorded exactly by means of another thread.

In the normal relaxed subject, it was observed that the tension rose only during the movement; the tension subsided when the limb came to rest in a new displaced position. In hemiplegia and Parkinsonism, however, such displacements were associated with a continued increase in tension.

[2]Using a stiff spring, the distortion corresponded to the tension to which the limb was subjected.

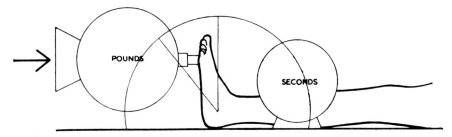

Fig. 2.2. A method that can be adopted by someone with no laboratory facilities. A balance, a clock and a cine camera was all that was needed. The interest was the fading of the tension needed to hold a certain position for 30s. (From Foley 1961, by permission.)

A more recent approach is that of Penn (1988). Rapid or slow stretches were obtained by pulling manually. The angle of the knee, the EMG and the force were monitored. In some observations, muscles have been stretched by moving the limb by hand while the EMG has been recorded.

Burke *et al.* (1971) used sinusoidal motion of the knee and monitored EMG activity in the quadriceps and hamstrings of cerebral-palsied patients. In both muscles there was a 'phase lead' of the electrical activity, indicating that the reflexes were not sensitive to the angle of displacement alone, but had significant velocity sensitivity. The phase lead was greater in the quadriceps. Another example of a rather similar approach has been used by Thilmann *et al.* (1991). Investigating the ankle in stroke patients they applied slow ramp stretches in the dorsiflexion direction, then in the plantarflexion direction, before finally returning to neutral. Force, position and EMG were monitored.

These methods are useful and within the reach of almost any laboratory, but are limited in that the rate of motion is only moderately repeatable and the relevant absolute parameters are unobtainable.

An interesting and perhaps unique variation of this system was devised by Foley (1961) (Fig. 2.2). The foot was dorsiflexed and the thrust recorded with scales and an electrically driven camera. The initial tension resulting from the displacement fell away and this was recorded over a period of 30s. This fading was investigated in children with spastic diplegia. Prolonged stretching gave a very real reduction of tone for about 10min. This curious and very interesting observation might well repay further investigation. In an inorganic system such behaviour would be known as 'stress-relaxation', a plastic property of matter, but perhaps in these children there was neural adaptation.

Tendon jerks

These responses are set up by the abrupt stretching of a muscle. In spite of their name, the receptors are the muscle spindles, not the tendon organs, and the contraction is generated reflexly. The proprioceptive fibres from the receptors conduct rapidly (*e.g.* 120m/s). The motor fibres from the spinal cord outwards are slower but still rapid (*e.g.* 60m/s) and the latency of the reflex is accordingly short.

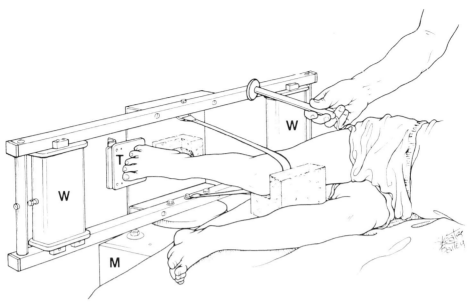

Fig. 2.3. Method of recording ankle jerk using a mechanical filter. A large motor in a housing (M) supports a beam at the end of which are fixed large weights (W), the ankle joint is concentric with the spindle, the ball of the foot is in contact with an isometric transducer (T) fixed to the beam. The motor provides a dorsiflexing force to put the calf muscles initially under a predetermined tension. The large weights scarcely move when the muscle contracts so the recording is effectively isometric (from Walsh *et al.* 1992).

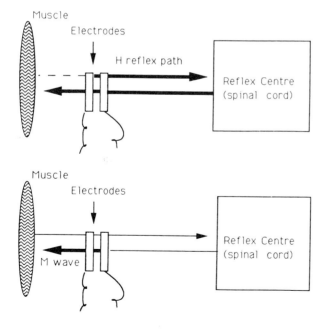

Fig. 2.4. Pathway for the Hoffmann reflex.

Fig. 2.5. Pathway for the direct muscle response.

However, as the muscle tension takes time to rise, by the time the muscle contracts the stretch has subsided.

The tendon jerks are an epiphenomenon; they serve in themselves no useful purpose but their existence indicates the existence of a pathway that is at times valuable in counteracting somewhat slower displacements—the stretch reflex. They enable the neurologist to ascertain the integrity of the system. In normal people, the briskness and ease with which they can be obtained vary greatly. Special importance is accordingly attached to any clear asymmetry.

Tendon jerks do not necessarily reflect muscle tone because they may be difficult to obtain when tone is high, although in moderate spasticity they are usually brisk. They are absent or weak in the long flexor of the thumb, even in subjects with brisk long-latency reflexes in this muscle. Again there is a lack of parallelism (Marsden *et al.* 1976*b*).

The muscular response also depends on thyroid status being slow in myxoedema (Lambert *et al.* 1951) and rapid in thyrotoxicosis. In the ankle jerk, the predominant effect is contraction of the soleus (Levy 1963), particularly if the knee is flexed, when the gastrocnemius (which acts over the knee as well as the ankle) is slack.

The timing of the events following a tendon tap offers a rapid and noninvasive way of obtaining some information about muscle properties. Unfortunately many workers have used isotonic systems which can only yield second-rate results. It is quite wrong to believe that the velocity of movement reflects the force generated by the muscle. If the system is isotonic, the swing will be influenced by the weight of the foot, the position of the centre of gravity, the inertia about the ankle and the passive elasticity of the dorsiflexors. These extraneous factors are essentially eliminated using an isometric technique. It is tiresome, however, if the initial load on the transducer depends on the weight of the foot, and the degree of relaxation for repeated rebalancing of the electrical circuit may be necessary.

It is better to start with the ankle subjected to a defined dorsiflexing torque whilst ensuring that the ensuing contraction is recorded isometrically. These requirements can be met if the patient lies on his or her side, with the sole of the foot in contact with a load cell mounted on a horizontal beam supported at its centre by a large printed motor (*e.g.* type G19M4). The motor is energised by direct current and applies a controllable torque. Heavy weights are fixed at the ends of the beam (Fig. 2.3). Because of this high inertia the contraction is almost over before there is any appreciable movement. Electronic circuits can give a digital reading of peak force, half-contraction time and half-relaxation time.

Electrical methods

A variety of different phenomena may follow the application of an electric shock to a nerve. The H reflex[3] is the reflex muscular response to a mild electrical shock to a

---

[3]The H refers to Paul Hoffmann, a physiology professor in Freiburg who used this technique in the early 1920s. It had been described by Piper a decade earlier.

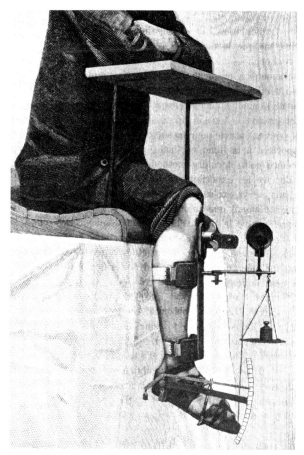

Fig. 2.6. The first myotonograph. The deflection of the foot due to the weight is read on the scale. Only one normal subject was used (from Mosso 1896a).

nerve, usually the medial popliteal (Fig. 2.4). The shock excites proprioceptive fibres, and hence activates the neural circuit responsible for the tendon reflex, but bypasses the muscle spindles. It is obtained using relatively weak shocks. Stronger stimuli give rise to an earlier M (for muscle) response: the motor nerve fibres are excited (Fig. 2.5).

In some measure, the H reflex mirrors motor neuron excitability. This is an important aspect of mechanisms responsible for abnormalities of tone but the procedure can be expected to monitor only phasic and not tonic aspects. There is a very large literature on the H reflex which has been excellently reviewed by Schieppati (1987).

Under certain circumstances, nerves other than those concerned with the calf muscles show an H reflex. Thus stimulation of the ulnar nerve in newborn infants gives a response in the hypothenar muscles. By about six months of age the reflex can no longer be found. However the pathways evidently persist, for it may reappear in Sydenham's chorea. Two patients with unilateral chorea studied by Hodes *et al.* (1962) showed the response only on the affected side. H reflexes in six patients with athetosis were monitored by Mizuno *et al.* (1971). They found an unusually early inhibition of the reflex, not found in normal people, arising from a weak shock to the peroneal nerve.

When motor nerve fibres are excited, impulses go not only down to the muscle but also upwards, antidromically, to the spinal cord. Sometimes the motor neurons are excited to respond to this and a second activation occurs, so that again impulses are sent to the muscle giving rise to the F response.

F stands for 'following': it follows the M response. The reaction can be useful for studies of conduction velocity. Thus if it is elicited by stimulating the ulnar nerve at the wrist, and if the M response in the same muscle is elicited by stimulating in the axilla, from the latencies, it is possible to estimate the conduction velocity of the motor nerves between the axilla and the spinal cord.

The F response may occur only once in every 20 occasions, so it may be necessary to use hundreds of trials to obtain a satisfactory average. Stronger shocks are needed than for the H reflex. The effect occurs more consistently following an upper motor neuron lesion and has been used to monitor the effects of dentatotomy for the relief of spasticity (Fox and Hitchcock 1982).

In one system, which surely could only have been devised in America, data about spasticity were transmitted by telemetry over the telephone to a central laboratory from the patients' homes. The aim was to monitor progress on drug therapy. The spontaneous EMG activity in the legs, that after a '100 volt non-painful electrical stimulus' *(sic)*, and that following dropping the legs, were sent over the lines. However, 'since electrical stimulation jams the telephone lines and distorts the polygraph record', it was omitted (Levine *et al.* 1972).

Gravity-driven methods

Mosso[4] (1896*a*) was evidently the first person to attempt to measure muscle tone (Fig. 2.6). The subject wore a sandal, and forces were applied to dorsiflex the foot by placing weights on a pan, the equipment being furnished with a cord and pulley. Alternatively, with a glass vessel on the pan, mercury could be poured so as to increase the force slowly. Mosso noted that the foot, having been displaced, did not in general return to its original position. Thus he came to these conclusions about the consistency of the musculature:

> Muscles are like lead and butter which, once they have been put out of shape, keep the mark they have received indefinitely.

The observation foreshadows later observations dealing with plastic and thixotropic effects. For the effects to be in scale one must think of quite a thin strip of lead or butter which is nearly melting! Traditionally neurologists have likened the resistance felt on manipulating the arm of a man with Parkinsonism to being similar

---

[4]Angelo Mosso (1846–1910), early experimentalist and prolific writer on physiological topics. He developed the ergograph for the investigation of muscular fatigue, and studied high-altitude physiology.

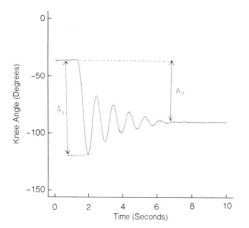

Fig. 2.7 *(above left)*. Wartenberg test in a normal subject. When the leg is dropped a number of swings occur. The relaxation index is 1.5 the value obtained by dividing A_1 by A_0. (Adapted from a figure kindly provided by Drs G.C. Leslie and N.J. Part, Department of Physiology, Dundee University.)

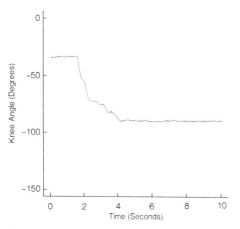

Fig. 2.8 *(above left)*. Wartenberg test in Parkinson's disease. The relaxation index is 1.0. (Figure kindly provided by Drs G.C. Leslie and N.J. Part, Department of Physiology, Dundee University.)

Fig. 2.9 *(right)*. Changes following a stroke. At the time of the record three weeks after the episode it is clear that the limb has become spastic. From top to bottom the relaxation index is: 1.8, 1.6 and 1.3. (Figure kindly provided by Drs G.C. Leslie and N.J. Part, Department of Physiology, Dundee University.)

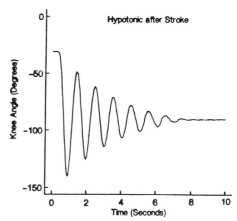
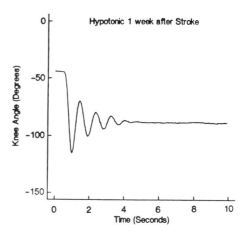
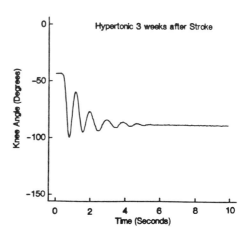

to that of bending a lead pipe—'lead pipe rigidity'. However the often-repeated dictum, that the limb in this condition remains where it has been passively placed, has not been confirmed by the analysis detailed in Chapter 10. When positioned on a suitable instrument, self-centring occurs: the wrist, for example, finding its own natural position.

The drive to obtain quantitative information has at times been motivated by the development of new therapeutic agents, the efficacy of which has been under investigation. Many people developed Parkinsonism following the pandemic of encephalitis lethargica in the early 1920s. Carmichael and Green (1928), interested in the efficacy of stramonium and other drugs that had been introduced, investigated the rigidity of the flexors and extensors of the elbow by an apparatus which was driven by a weight, the motion being recorded on a smoked drum. The time taken for the forearm to fall a certain distance was measured, longer times implying greater ridigity. The introduction of a new drug, diazepam (Valium, a benzodiazepine) was the impetus for holding a symposium on skeletal muscle hypertonia. The discussions covered a number of different measurement procedures (Levine 1964). Another useful account of methodology is that of Haley and Inacio (1990).

An early attempt to obtain absolute parameters relating to the tonus of skeletal muscle in man was that of Smith *et al.* (1930). They allowed the leg to drop and recorded the fall. For the dynamical analysis they needed to know the weight of the leg, its moment of inertia about the hip, and the position of the centre of gravity. None of these parameters could be obtained from their method, and the study can only be described as a glorious failure. Two German workers in 1889 had made measurements on dismembered frozen corpses (p. 68). It was unsatisfactory for Smith and colleagues to extrapolate from this data, because there are very wide variations in the moment of inertia between different subjects (see Chapter 6).

A popular method of assessing tone uses instrumentation to quantify the Wartenberg test (p. 8), and Bajd and his colleagues in what was Yugoslavia have made extensive observations (Bajd and Vodovnik 1984). In spasticity there is usually a brisk and early retardation of the drop of the leg. Smith *et al.* (1930) had noted that the fall is often checked so vigorously that the limb actually rises. The falls were by jerks with alternating contractions and relaxations. A 'relaxation index' has been used in the more recent analyses. This is the ratio between the amplitude of the first swing and the total movement from start to end. In normal people the leg first swings well past the vertical so the relaxation index is well above one (Fig. 2.7). In severe Parkinsonism the leg falls to its final position without oscillating. The relaxation index is thus about 1 (Fig. 2.8). In spasticity, because of the check usually found soon after the fall starts, the relaxation index is less than 1. Records showing the development of spasticity following a stroke are shown in Figure 2.9. The differences in these indices in spasticity and rigidity have been documented by Brown *et al.* (1988).

The merit of this method is its relative simplicity. It has been used to evaluate

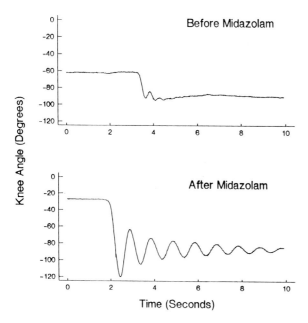

Fig. 2.10. Intrathecal administration of a 2mg of midazolam, a benzodiazepine preparation, for spasticity due to multiple sclerosis. The procedure was adopted to evaluate the possible effects of intrathecal phenol to relieve spasticity on a more lasting basis. The record was taken 35 min after the injection. (Figure kindly provided by Drs G.C. Leslie and N.J. Part, Department of Physiology, Dundee University.)

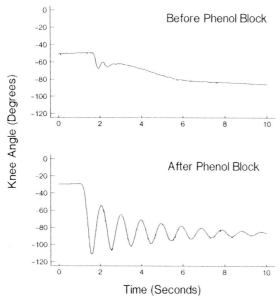

Fig. 2.11. Effect of intrathecal phenol in relieving spasticity in a patient with multiple sclerosis. (Figure kindly provided by Drs G.C. Leslie and N.J. Part, Department of Physiology, Dundee University.)

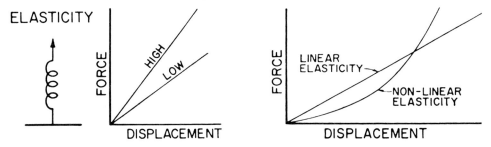

Fig. 2.12. *Left:* elastic stiffness as exemplified by an ideal spring showing a linear relationship between force and displacement: the higher the slope the stiffer the spring. *Right:* example of non-linear stiffness, the slope increasing with displacement. (From Wright and Johns 1960, by permission.)

the effects of blocking the spinal roots by intrathecal agents (Figs 2.10, 2.11). A useful evaluation of intrathecal chemotherapy is by Kasdon and Abramovitz (1990).

The acute effects of applying phenol to the dorsal spinal roots of cats was studied by Iggo and Walsh (1960). Using natural stimuli, proprioceptive fibres were more resistant than cutaneous fibres. When the compound action potential was evoked by an electric shock in a frog's sciatic nerve, the slower fibres were the more susceptible. The damage caused on a long-term basis, however, does not follow this pattern because fibres of all sizes are destroyed. Glenn (1990) reviewed the use of nerve blocks for the relief of spasticity.

Nathan (1968) investigated the effects of the intrathecal injection of a local anaesthetic in patients with spastic paraparesis. A variety of stimuli (tapping the knee jerk, cooling the skin with ethyl chloride, pricking, scratching or pinching) set off a reflex not seen before the injection. This consisted of bilateral flexion abduction movements of both legs. Surprisingly, in some of the patients excessive muscle tone was reduced for days following the procedure, though one would suppose that the drug itself would be cleared away in a matter of hours. This interesting finding remains unexplained.

Discussing methods of measuring tone, Fenn and Garvey (1934) noted that:

> For a scientific analysis each method must measure some one clearly defined physical property of the muscle in *absolute* units. Just as surely as a muscle has length and breadth it has elasticity and viscosity, etc. which may show functional changes, and these properties must be measured quantitatively, not qualitatively, in proper physical units.

Judged on this basis, the gravity-driven methods are failures.

An engineering approach
It is important to consider some of the basic principles of mechanics. In particular the questions of stiffness, viscosity and inertia.

Stiffness
An elastic substance is one in which stress, force, is a function of the strain, displacement. The relationship may be linear as in an ideal spring, or non-linear as in a rubber band (Fig. 2.12).

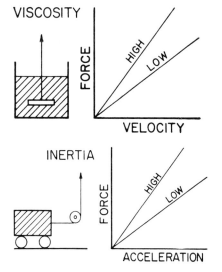

Fig. 2.13. Viscous friction as exemplified by a plate moving through an 'ideal' viscous fluid showing the relationship between force and velocity. Increased viscosity results in a steeper slope. (From Wright and Johns 1960, by permission.)

Fig. 2.14. Inertial stiffness as exemplified by a mass on frictionless bearings. (From Wright and Johns 1960, by permission.)

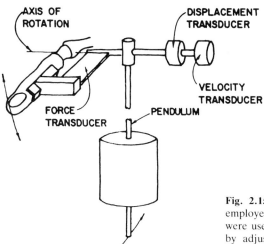

Fig. 2.15. Swings of up to 30° of the finger were employed with this instrument. Only low frequencies were used (0.6 to 1Hz). The variation was achieved by adjusting the length of the pendulum. (From Wright and Johns 1960, by permission.)

In this work the concern is with motion at a joint and hence with rotatory, not linear, stiffness. The concept of torque is relevant (Latin *torquere*, 'to twist'). It is defined in the Oxford English Dictionary as 'the twisting or rotatory force in a piece of mechanism'. The numerical value is given by the force applied multiplied by its distance from the axis. The units are thus N m. In this book sometimes for convenience the term 'force' will be used when 'torque' would be more precise, but the context will make the meaning clear.

For rotary stiffness the units are newton metres per radian (N m/rad).

The term 'stiff' then has this exact meaning. It is also used loosely by non-technical people to denote any resistance to motion. In this book it is used in its strict sense. It is unfortunate that some scientists are not precise in their terminology. In some experiments on isolated muscle fibres it has been said that stiffness depends on the rate of stretch. This is nonsense. A consideration of the records shows that the fibre is exhibiting viscous as well as elastic properties. Otherwise admirable work, relating muscle mechanics to biochemical events, is reviewed by Brenner (1990).

Viscosity

A viscous substance is one in which the stress (torque) is a function of the velocity. The relationship may be linear as exemplified by a plate moving through an ideal viscous liquid (Fig. 2.13). For rotary motion the units are N m/rad s^{-1}. Viscosity is sometimes referred to as 'drag'.

The word 'viscous' comes from the Latin *viscus*, meaning 'mistletoe'. The juice of the berry is sticky and was used to trap birds. If they landed on a surface on which 'bird-lime' had been spread, their feet were entangled and they could not take off.

The parts of the living body are lubricated with slimy liquid. As a muscle moves, its coverings slip on adjacent structures, but not without an impediment to the motion. Static friction, so conspicuous and indeed often essential with dry inanimate objects, is notably absent in the living body.

Inertia

With a mass moving on frictionless bearings, the force producing motion is a linear function of the acceleration (Fig. 2.14). For rotary motion the relationship is between the moment of inertia and angular acceleration. The units are kg m^2; where these are inconveniently large, as in certain instances in this book, the values are given as g m^2.

One use of inertia was in the flywheels of steam traction engines. Because much force is needed to change the rate of rotation of a heavy mass, the effect was to damp out fluctuations.

Contributions of elastic stiffness, viscosity, inertia, and friction to the resistance to motion

A refreshingly original study was that of Wright and Johns (1960). In their apparatus a heavy pendulum could cause a finger to oscillate at 1Hz or a little slower (Fig. 2.15). The displacement, the velocity of the movement and the force imparted to the finger were recorded. The intention was to study the stiffness of the joint but the motion would naturally have involved the muscles, and the contribution of the joint in such experiments is discussed later (p. 89).

The interest of this study is that the analysis was in terms of the various physical properties of the tissues. As these concepts will be significant in later sections of this book, they will be discussed in some detail. The type of record obtained is shown in Figure 2.16. There is a hysteresis loop. The authors analysed

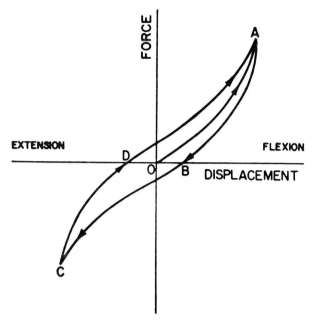

Fig. 2.16. The type of record obtained on using the apparatus shown above. There is hysteresis. (From Wright and Johns 1960, by permission.)

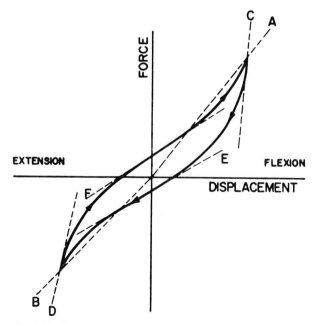

Fig. 2.17. The use of tangents to the curve to estimate stiffness. The conclusions are, however, highly questionable since detailed observations, discussed later in this book, have repeatedly shown that the postural system is stiffer for small than larger movements. (From Wright and Johns 1960, by permission.)

their graphs as shown in Figure 2.17. By taking the slope of the tangents, they assumed they were measuring elastic stiffness. Thixotropic effects were not known at this time (see Chapter 7). At the turning points, where velocity falls away to nothing, thixotropic stiffening is maximal. The belief that these tangents adequately measured elastic stiffness is hopelessly flawed.

For a further review of this work it is necessary to consider some mathematical relationships. For an oscillation of peak amplitude A and frequency f the values at any instant t are:

$$\text{Displacement} = A \sin 2\pi f t \qquad \text{(equation 5)}$$
$$\text{Velocity} = 2\pi f A \cos 2\pi f t \qquad \text{(equation 6)}$$

The authors believed, surely correctly, that static friction was negligible: but their conclusion was that elastic stiffness was the dominant constraint. This should not be accepted at face value, for the low frequency of the oscillation will have involved low velocities. If higher rates had been employed, the contribution of the elastic component would have been unchanged but that of the viscous component would have increased. If the frequency had been 10 times as high the viscous resistance will have been 10 times as great for the same amplitude of swing (equation 6).

The authors estimated inertia from the dimensions and the assumed density of the finger, and considered that this factor contributed to a negligible degree to the resistance to motion. This is not satisfactory, because in moving the finger a large mass of muscle too will be shifted. The estimate of inertia is almost certainly far too low. Furthermore the frequency used will have an overwhelmingly large effect. The relevant equation is:

$$\text{Acceleration} = 4\pi^2 f^2 A \sin 2\pi f t \qquad \text{(equation 7)}$$

For an oscillation at 10Hz the acceleration, and hence inertial resistance, will be a hundred times greater than the corresponding movement at 1Hz.

Summary

The engineering approach is fundamentally better than many of the methods described previously; but for a satisfactory evaluation of the resistance to motion of a limb, it is necessary to go to at least a medium level of technology.

3
DISPLACEMENT AS THE INDEPENDENT VARIABLE

In more advanced instruments than those discussed earlier there are two main choices. While the relationship between torque (stress) and the movement (strain) is the essence of the problem, torque may be the dependent or independent variable. The choice lies between inducing a certain predetermined rate of displacement (stretching) or subjecting the limb to a certain defined variation of force (pulling). The choice may seem to be of little consequence, but this emphatically is not so. Very different sets of data are obtained according to the selection. In experiments on decerebrate muscle tone in cats, Roberts (1963) made this distinction quite clearly. The implications have frequently been overlooked.

Techniques
Rhythmic isokinetic methods
Possibly the first example of a myotonograph where the velocity of stretching was closely controlled was that devised by Fenn and Garvey (1934) (Fig. 3.1). Like a number of other similar instruments, it used a geared motor to produce the motion. The motor was strong enough (¼ horsepower) to overcome any resistance of a relaxed limb which was moved to and fro. The experimenters were correct to position the instrument so that the oscillations were in the horizontal plane. To use vertical movements, as some investigators have done, is quite unsatisfactory because the force due to gravity will vary with the angle, and hopelessly compromise any calculations. It was arranged that the motion, alternately into flexion or extension, was at constant speed (*i.e.* 'isokinetic'). The velocity reversed abruptly at the ends of the travel. The purpose of this system was to eliminate forces during the motion due to inertia, but there were naturally large excursions of force at the turning points. Measurements were made at a number of different speeds and the recorded force increased in an approximately linear manner according to the selected velocity.

A similar device but with more advanced technology was used extensively by Webster (1964, 1966) to evaluate Parkinsonian rigidity and spasticity. He developed instruments to test both the elbow and the knee, 'softened' the turning points to reduce the jolt of an abrupt change of direction, and measured the work done by the motor during the to-and-fro cycles in moving the leg.

Unidirectional isokinetic methods
In some instruments, motion in only one direction has been employed instead of alternating movements. An example of such an apparatus is shown in Figure 3.2.

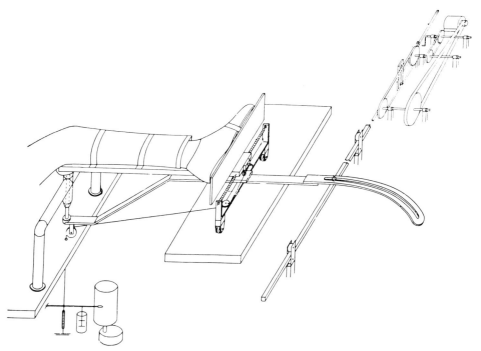

Fig. 3.1. An electric motor (top right-hand corner) drove a chain carrying a toggle drove a lever system and a to-fro motion at the knee was obtained. There was an abrupt reversal at each end of the travel which was otherwise at constant velocity. There was considerable vibration. The force generated, determined by recording the distortion of a spring, was recorded on a rotating smoked drum. Not drawn to scale, the motor and driving gear were larger than portrayed. One record was obtained of a patient who readily went into a tetany on overbreathing. The resistance to the motion greatly increased and finally became so great that the motor was unable to move the leg at all. (From Fenn and Garvey 1934, by permission.)

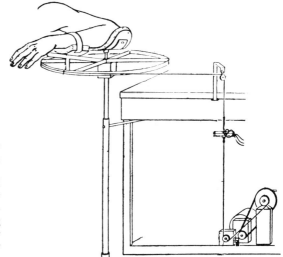

Fig. 3.2. In this system a motor drove the arm, at constant velocity, from flexion into extension through a cable and drum. The tension in the cable was recorded electrically. The movement was very slow (90° in 23s). There is much more likelihood of voluntary activity interfering with the measurements during slow movements. (From Agate and Doshay 1956, by permission.)

Doshay (1964) pursued the assessment of hypertonus doggedly for 30 years, but the effort was not without its disappointments:

> The rigidometer is a precise measuring instrument . . . but requires highly complicated calculations including calculus for an exact determination of the change in hypertonicity. In the course of years, Professor Agate made successive modifications in the machine, including an abortive effort to adapt it to leg measurements. My original hope had been to achieve an apparatus that would quickly tell the investigator that a new drug 'X' is perhaps 37 per cent more, or 20 per cent less, effective than Artane. However, by the time of my retirement in 1962, this goal had not been reached, although with or without the benefit of the rigidometer, we did succeed in evaluating some 200 new synthetic compounds, most of which proved unsuitable for Parkinson's disease, or were too toxic.

Replication of results is more cumbersome with this type of instrument, which must therefore be regarded as inferior. Nevertheless, some interesting results in patients with spasticity were obtained by Knutsson (1985):

> At the slowest speed, 30 deg/s, the restraint was so low that it can not readily constitute any impediment to function. With increasing speed, there was a successive increase in spastic restraint. At a speed of 120 deg/s, it gave an opposing torque of 15 Nm. This speed of knee flexion is only about half that used in free speed walking in healthy man. Thus, the spastic restraint could be expected to limit the capacity for fast movements . . . In some patients, there is a strong resistance even at low speeds. In these, the resistance at higher speeds may become so large that it can not be measured without risk of injury.

This factor of speed of movement is obvious on the clinical examination of many patients with spasticity. The point was also made by Burke (1980):

> The characteristic feature of the stretch reflex in the spastic patient is that it is velocity-dependent. If a spastic muscle is stretched slowly there may be no reflex contraction and no detectable increase in muscle tone, but a reflex response becomes apparent with faster stretching movements and builds up approximately linearly with increases in the angular velocity of the stretching movement. The prominent velocity dependence of the spastic reflex probably results from the dynamic sensitivity of the primary spindle ending. This ending is, after all, the only muscle stretch receptor with significant velocity sensitivity.

Further isokinetic studies were reviewed by Haley and Inacio (1990). Some investigators have used the expensive commercial systems popular in sports medicine.

Sinusoidal methods
Sinusoidal motion of fixed frequency was employed by Thompson *et al.* (1978). A stepping motor was the power source. The standard test used motion of 10°, peak to peak, with a frequency of 0.1Hz. The velocities were thus very low. The cycles were applied around mid-positions at 10° increments between 50° and 10° of flexion. They were interested in the changes that may occur when joints are arthritic. Sinusoidal motion of a finger was used by Yung *et al.* (1984). The frequency was 0.1Hz and the peak-to-peak amplitude 8°; the motion was thus again small and very slow. According to these workers there is a diurnal variation of resistive torque, but I have reservations about this interpretation. Stiffness was maximal at 2 a.m., but can anyone hauled out of bed at such a time be properly relaxed?

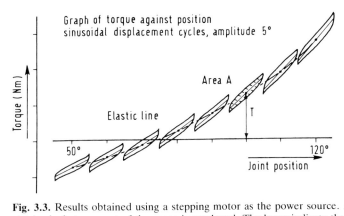

Fig. 3.3. Results obtained using a stepping motor as the power source. A series of hysteresis loops is obtained whatever part of the range is employed. The loops indicate that energy is being absorbed by the movement—how far this is due to viscosity and how far to thixotropy is not clear. (From Thompson *et al.* 1978, by permission.)

In the method of Duggan and McLellan (1973), too, the imparted motion was sinusoidal but the frequency of operation could be varied.

In an approach that was highly sophisticated mathematically, Lehmann *et al.* (1989) used a movement of 2.5° on either side of the mid-position at frequencies up to 12Hz. They measured not only the torque resulting from the rhythmic displacements but also the phase angle. They could thus partition the developed force according to its relationships with the motion. The impedance 'in quadrature' with displacement (*i.e.* the value in step with velocity) was taken to represent viscosity (see below). The component of impedance in step with displacement was regarded as consisting of a value dependent on stiffness, independent of frequency, and a value, of opposite sign, which varied with the square of the frequency and was proportional to inertia. By ascertaining the change as the frequency was altered, an approach could be made to estimating both stiffness and inertia. The shortcoming of this approach lies in the non-linearity of the system and the presence of significant plastic effects.

Significance of the results
Measurements of viscosity?
These systems are restricted to relatively low frequencies of operation or quite small movements. For rapid oscillations of any significant amplitude the inertial forces would become intolerable and potentially damaging. To contain the inertial effects would require complex engineering; to keep them constant as the frequencies were changed would involve a complicated instrument for the displacement would have to be inversely related to the square of the frequency.

Fenn and Garvey (1934) believed that their method measured viscosity. This view requires some qualification. Because of the low rates of motion, thixotropic effects may be significant or even the dominant source of resistance to motion. These phenomena are discussed in Chapter 7. They are doubtless the reason why the tension curves begin with a steep initial rising phase particularly after a pause when the limb has been motionless (Hufschmidt and Schwaller 1987). Deductions about viscosity may possibly be satisfactory, but the argument is 'messy'.

In clinical spasticity high values for viscosity include 'reflex viscosity', the force developed as a result of proprioceptive reflexes initiated by velocity-sensitive reflexes.

Measurements of elastic stiffness?
The records of resultant torque plotted against position have invariably been loops. Examples are shown in Figure 3.3, which may be compared with Figures 2.16 and 2.17.

The interpretation of these loops may be considered in the light of the classical Carnot cycle of a heat engine. On the length tension diagram the area under the upper curve represents the integrated product of force times distance and thus the energy delivered by the motor. The area under the lower curve represents the energy absorbed during the return movement. The difference in the areas represents the net energy delivered to the limb during a full cycle.

At the turning points, where velocity drops to zero, the contribution of viscosity must be zero, but with the isokinetic methods the inertial forces are massive and it is thus not appropriate to take the slope of the line joining the ends of the loop as representing stiffness. With the sinusoidal method both the inertial and the viscous forces are making a constantly varying contribution to the resistance to motion so the slope of the loops is compounded of different effects.

A physician who only manipulated a limb through a very small distance would not learn much about its properties; s/he would be regarded as a woefully incompetent clinician. An instrumental method which depends solely on quite small movements can be expected to yield restricted information; while if the motion is more substantial but slow, the data about dynamic responses will be inadequate. Not every method described is valuable. There are examples in the literature where it might be much better to depend on clinical testing.

Deduction

It is important for things to be done the right way round. The wheel of a barrow may turn on the shaft (wrong way), or be fixed to the shaft which turns in bearings (right way). One way and another, these stretching methods are hamstrung[1].

The conclusion is that in spite of their apparent attractiveness, ease of construction, the commercial availability of isokinetic exercisers, relative simplicity, and 'scientific' appearance, the stretching methods have proved to be very poor analytical tools. The alternative—pulling—is the subject of the next chapter.

---

[1]This is another expression based on a muscular metaphor. The word refers to cutting the tendons of a man, or horse behind the knee. The Oxford English Dictionary quotes William Youatt: 'The Israelites were commanded to hough or hamstring the horses that were taken in war'. A man who is hamstrung is unable actively to flex the knee and so cannot run; he can still walk, however, because the knee will flex passively as the thigh moves forward.

4
TORQUE AS THE INDEPENDENT VARIABLE

Some 35 years ago the author was invited by a neurosurgical colleague, Professor J.G. Gillingham, to measure muscle tone in patients with Parkinsonism being prepared for stereotaxic operations. No suitable method seemed to be available, and many years later it was decided to try a new approach. Bearing in mind the findings of Roberts (1963), it seemed that interesting results might be obtained by applying rhythmic forces to the limb. The procedure did not appear to have been tried before as a method of assessing tone in man, and a search for a suitable technique was instituted.

Technology
Preliminary studies
An initial approach was to build a small instrument with a miniature electric motor of adjustable speed that drove, through a bevel gear, in opposite directions, two wheels facing one another (Fig. 4.1) (Walsh 1968*b*). The wheels carried eccentric weights so that a sinusoidal force was generated when the instrument was in action. This arrangement was used to generate force in a linear direction. Had a single rotating eccentric mass been used, the developed force would have been rotary. Through a second bevel gear, the motor shaft also drove another similar pair of wheels furnished with eccentric weights. This pair was mounted in a cage that could be rotated slowly by the smallest available electric motor. According to the position of the cage, the forces from the two pairs of eccentric weights added vectorially and so could be varied from zero to double that which would have been provided by a single pair of weights alone. The frequency and strength of the rhythmic force was thus adjustable by feeding appropriate currents to the two motors.

As the instrument produced a torque related to the square of the revolution rate, and as the motion was recorded by an accelerometer, it was of limited use at low frequencies. Although efforts had been made to construct an instrument that was as light as possible, the weight was not negligible and this added considerably to the inertia. Both in normal subjects and in children suffering from cerebral palsy, however, the motion was greatest at a certain rate of operation. This finding foreshadowed later work on limb resonances.

Because the foregoing apparatus loaded the hand, another approach was used (Fig. 4.2). The differential gear from a small petrol-driven three-wheeler, a 'Harper' invalid carriage, was obtained, and eccentrically mounted winding drums were fixed on the two opposing shafts. A variable speed motor drove the mechanism through one of these, and the two pulleys thus rotated in opposite directions. Around each was wrapped 2½ turns of nylon cord.

The drums acted as capstans and the cords transmitted rhythmic forces to a

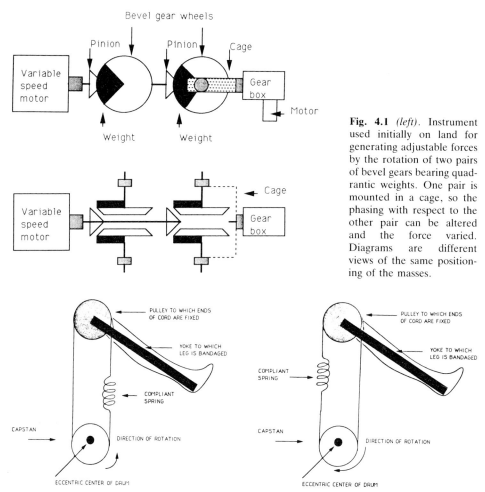

Fig. 4.1 *(left).* Instrument used initially on land for generating adjustable forces by the rotation of two pairs of bevel gears bearing quadrantic weights. One pair is mounted in a cage, so the phasing with respect to the other pair can be altered and the force varied. Diagrams are different views of the same positioning of the masses.

Fig. 4.2. View from the two sides of the arrangement in which rhythmic forces were generated by a capstan arrangement.

second set of pulleys mounted on dead bearings on either side of the knee. The capstan action amplified the tension produced in the cords by light springs and this varied according to the rotation of the eccentric surfaces. The second set of pulleys was coupled to the leg by a yoke. The shaft that had been driven by the petrol motor, when in use in the vehicle, could be fixed or rotated slowly. This adjustment of the gearing changed the phasing of the two pulleys. According to this phasing, the forces added or opposed one another and the resultant torque on the joint could thus be adjusted from zero upwards to the maximum which the instrument could provide.

Another system employed by Berthoz and Metral (1970) used a magnetic clutch. A pulley driven at constant speed was coupled to a second by

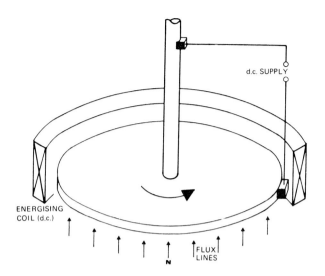

Fig. 4.3. The Faraday disc dating from 1831. The disc lies in a magnetic field. A current passed through the disc causes it to rotate. The principle is used in all electric motors, a conductor carrying current experiences a force if in a magnetic field. (From Evans 1972, courtesy of ETI Electronics.)

electromagnetic power activated by a magnetic field that was proportional to the current from a function generator. The system was used to test not resting muscle tone, but the postural stability of the forearm held vertically.

Printed motors

The printed circuit motor[1] was invented in 1958 by F.H. Raymond and J. Henry-Baudot at the *Société d'Electronique et d'Automatisme* in Paris (Knights 1975). It is a basic fact of electromagnetism that a current flowing in a magnetic field produces a force. This is the principle behind the disc invented in 1831 by Faraday (Fig. 4.3). Unfortunately, because in essence the armature consisted of a single turn, it required the supply of very high currents at low voltages. This was unwieldy, consumed a lot of power, and required the brushes to pass very heavy currents. In the printed motor the armature is made in a thin pancake (Fig. 4.4). The flux air gap is reduced to a minimum and no iron is required. In conventional motors, torque modulation results from the reluctance changes as the slots of the commutator pass under each pole tip and can be troublesome at low speeds. This undesirable cogging characteristic is absent in the printed motor. The end connectors between conductors are arranged geometrically to provide a running surface for the brushes which deliver the supply power to the motor, thus dispensing with the need for a separate commutator. The use of such a device to generate torques for biomechanical studies enables the development of a much more versatile system:

---

[1] The term 'printed' refers to the technology initially used to manufacture the pancake, the method being similar to that for fabricating printed circuits for electronics. These motors have many uses; one has been to move tape in an earlier generation of computers where they have been useful because of their rapid response. Most of the motors used were obtained from Printed Motors Ltd, Oakhanger Road, Bordon, Hampshire GU35 9HY.

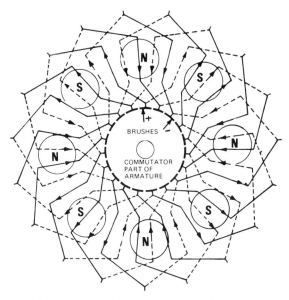

Fig. 4.4. The armature of a printed motor in simplified diagrammatic view. As there is no iron there is no attraction to the stationary parts of the machine. The absence of this attraction and the light weight of the armature means that the motor bearings operate under lightly loaded conditions and the motor generates less frame conducted noise than other types of electrical machines. By alternating the polarity of adjacent pairs of poles the conductors under each pole can be connected in series and it is thus possible to design for practical levels of voltage keeping current to manageable proportions. For simplicity only a small number of conductors per pole have been shown. (From Evans 1972, courtesy of ETI Electronics.)

1. The mass of the moving parts is small, so that the inertia added by coupling the driving element to a limb can be trivial.

2. The low inductance of the armature, typically 50μH, gives a short electrical time constant. This allows the current to rise very rapidly. Changes of torque can be developed almost instantaneously; when required, the force on a limb can be reversed abruptly. The force on the limb can thus follow the input from a waveform generator, giving great flexibility in use.

3. Provided there is no damage by mechanical failure, such a motor can sustain a heavy overload: *i.e.* heavy currents can be used for short periods, which if continued would cause unacceptable heating. The torque available is not limited by saturation as in a conventional motor, because there is no magnetic material in the armature. On a short-term basis, or using a small duty cycle, a small motor can thus generate substantial force. There is a wide range of sizes available; some are furnished with special magnets and thus can generate more force for a given current. The suffix H is used to indicate such a high torque motor.

4. The torque produced by the motor is linearly related to the current passed. A convenient and simple method to monitor the applied torque is to record the voltage drop in the earth return lead from the motor. A separate transducer is not needed.

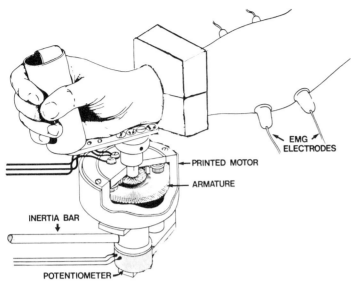

Fig. 4.5. The apparatus used for many of the observations recorded in this book. The printed motor is portrayed in cut-away form; it is concentric with the wrist joint. The fingers are held in contact with the handle by a strap. The lower end of the forearm is restrained by the two halves of a shaped block of plastic. Suction cup electrodes are used for the EMG recordings and the motion is recorded by a potentiometer. The inertia bar, used only for occasional studies such as those described on pp. 69–71.

A new motor is likely to have unacceptably high friction. This can be reduced by an extensive period of 'running in' and by degreasing the bearings. One of the most satisfactory motors (type G16M4) was obtained on the surplus market. Motors can successfully be modified to improve their performance for physiological purposes; there is an illustration of such a modified motor in a paper by Marsden *et al.* (1976*a*):

> Friction was reduced by removing two of the four brushes, filing down the remaining pair to diminish the area of contact and reducing the brush pressure. A spring cup washer which caused extra friction was removed from the spindle . . . To increase the torque available the light alloy member forming the body of the motor was replaced by four brass spacing pieces so that air could be blown through to cool the armature and allow the use of larger currents.

Remagnetisation is required after a motor has been opened. This is accomplished by passing a very hefty current for a very short time through the wires wrapped round the magnets.

Apparatus for testing the wrist
An important aspect of the clinical examination of patients with Parkinsonism involves manual testing for rigidity at the wrist. As the system was initially intended for investigating this clinical condition it was decided to develop a system for measurements at this joint based on a suitable motor. Type G9M4 was considered to be of the appropriate scale (Fig. 4.5). The motor has a double-ended shaft. One

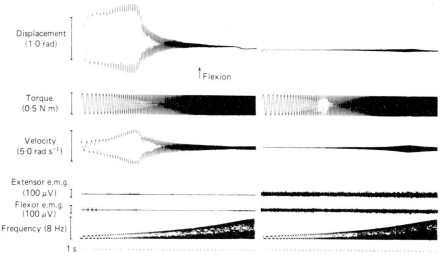

Fig. 4.6. Resonance at the wrist with a subject relaxed (*left*) and with moderate co-contraction of the flexor and extensor muscles. The resonant frequency rises (from Lakie *et al.* 1984*b*).

end is furnished with a light alloy handle of adjustable length; the joint is concentric with the motor and the fingers are held with a Velcro strap. The other end communicates via a boss with a low-friction conductive plastic potentiometer to record displacement. In a later version of the apparatus an optical potentiometer[2] was used; such a device has 'infinite resolution, no friction and no noise'. The boss is furnished with a transverse hole through which round metal bars can be inserted, so that the effects of adding inertia to the system can be studied. By using bars of different lengths, varying amounts of added inertia can be employed. A signal corresponding to angular velocity is obtained by passing the output of the potentiometer through a pseudo-differentiator. The EMG of the flexor and that of the extensor muscles are obtained by the use of suction-cup electrodes. By these various means it is thus possible to record, on a pen writer, the applied torque, angular position, angular velocity and the EMG signals.

The motor was energised from a power amplifier using feedback so that the current through the motor faithfully followed the various waveforms, delivered to the input, from a custom-built waveform generator. Sinusoidal torques could be generated of fixed frequency, or they could be programmed to rise in rate, exponentially producing what in engineering parlance is called a 'chirp'. As the wrist is not lightly damped, it is acceptable to use a relatively rapid rate of sweeping; a sweep rate of 0.33 octave/s^{-1}, however, is too rapid to get a smooth resonance curve (Walsh 1975*a,b*).

Resonance at the wrist

As soon as the apparatus was set up it was clear that the wrist resonated. As the

---

[2]Precision Varionics Ltd, Units 8 and 9, Cmw Draw, Ebbw Vale, Gwent NP3 5AE.

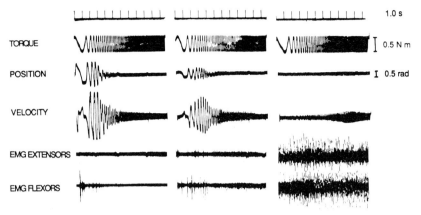

Fig. 4.7. Resonance at the wrist in a subject relaxed (*left*), slightly stiff (*centre*) and very stiff (*right*). Another example of an increase of resonant frequency with voluntary stiffening.

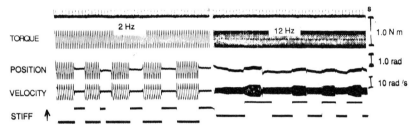

Fig. 4.8. Effect of voluntary stiffening at a low frequency (2Hz), contrasted with the effects at a comparatively high frequency (12Hz). The motion is virtually abolished at 2Hz but is accentuated at 12Hz; the stiffening has retuned the system so that it is at, or close to resonance, at the higher frequency.

frequency of the torque rose the oscillations increased to a maximum (the resonant frequency). With further increase the motion fell off again. The resonant frequency varied with the state of the muscles, being higher when there was voluntary stiffening (Figs 4.6, 4.7). Accordingly it was felt that the system would probably be useful for monitoring pathological conditions giving rise to muscular hypertonia.

It was noted that at low frequencies the motion was greatly reduced by voluntary stiffening, but the converse could be found if an appropriately adjusted higher frequency was employed (Fig. 4.8). The forearm had become tuned to a higher frequency. The hand and associated structures have mass and the muscles have stiffness. With the stiffening the mass was unchanged but the muscle stiffness changed—this was reflected in the records. The arrangements were clarified by models (Fig. 4.9). The system was essentially a torsion pendulum and the behaviour of such a system is well described in textbooks of engineering. The relevant equation is:

$$f_0 = \frac{1}{2\pi}\sqrt{\frac{K}{J}} \qquad \text{(equation 8)}$$

where f_0 is the resonant frequency, K the stiffness and J the inertia. It is thus

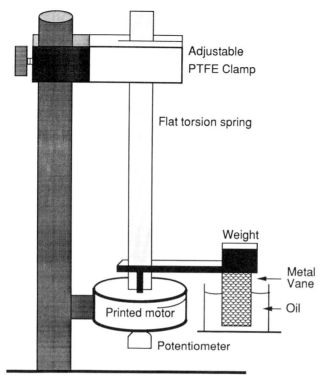

Fig. 4.9. Model for demonstrating principles of resonance. The printed motor is coupled to a flat spring which can be clamped at different distances thus varying the stiffness. Inertia is provided by a weight fixed to a lever. Damping is provided by a vane moving in oil. In a more sophisticated version, negative velocity feedback, instead of the oil, provides the damping. This allows one to adjust the damping in a more versatile manner.

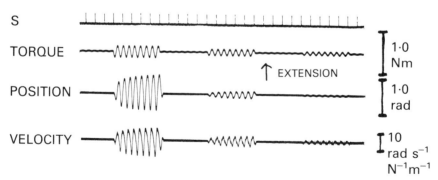

Fig. 4.10. Wrist is oscillated at a low frequency. Force is halved successively, motion drops away disproportionally showing non-linearity; it is stiffer for the small forces. The amplification of the velocity trace is automatically increased as the force is reduced, the envelope accordingly reflects the 'admittance' of the system (Reciprocal of Mechanical Impedance).

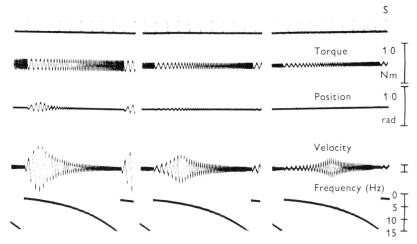

Fig. 4.11. Wrist is supplied with chirps at three levels. The response shows stiffening with the smaller forces for the motion falls off non-linearly and the resonant frequency rises. Velocity calibration 1 rad s^{-1} at the highest force, 0.5 rad s^{-1} for the middle record and 0.25 rad s^{-1} for the right-hand trace (from Walsh 1973a).

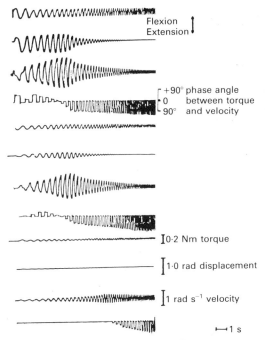

Fig. 4.12. Resonance at the wrist at three torque levels. A phase detector was used. At a zero crossing of the velocity signal, once per cycle, a triangular wave was passed to a sample/hold circuit. From the triangular wave the driving sinusoid was derived in the waveform generator. Below resonance velocity leads torque, at resonance it is in phase with torque and above resonance velocity lags behind torque. When the velocity is very low, as it was with the lower torques below resonance, the detector does not function properly and artefacts have been deleted.

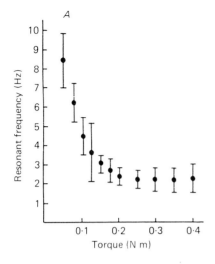

Fig. 4.13. Resonant frequency of the wrist is plotted at 11 peak torque levels. The results are the mean values (± S.D.) of the right wrists of 12 male subjects. For small torques the system is non-linear, the resonant frequency falling as the torque rises becoming constant at about 0.22 N m (from Lakie *et al.* 1984*b*).

apparent that it is the square of the resonant frequency that reflects the stiffness. Strictly speaking the equation is only applicable when damping is negligible, but the error is quite small for the situations being described in this chapter, and has accordingly been neglected.

When a sinusoidal force of fixed and low frequency was used it was observed that the motion was not linearly related to the applied force. Proportionally there was less movement for the smaller forces (Fig. 4.10). If the limb had behaved linearly and Hooke's law[3] had been obeyed, resonance would have been the same when different force levels were tried; but in fact for the smaller forces used the resonance was at a higher frequency (Fig. 4.11). On halving the torque the resonant frequency fell, and on halving it again there was a further fall. At low frequencies the torque was in phase with displacement, at resonance it was in phase with velocity, and at frequencies well above resonance it was in antiphase with displacement. The result of using an electronic circuit for displaying phase is shown in Figure 4.12.

A systematic study was undertaken of the relationship of the resonant frequency to the applied torque (Fig. 4.13). It will be seen that the resonant frequency is sensibly constant for the medium and higher torques, but that below a certain critical level it rises progressively. The reasons for this increase are discussed later (Chapter 7). A lower critical level is found in women (Fig. 4.14), presumably because they are generally less muscular (p. 75).

For very small forces the wrist resonates at about 9Hz, which has a bearing on the mechanisms underlying physiological tremor (pp. 94–98).

Neural influences?

To what extent are these effects due to neural effects such as activity in the stretch

---

[3] Hooke's law is discussed on p. 172.

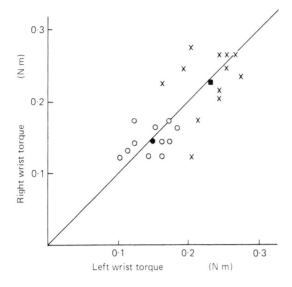

Fig. 4.14. Critical torque levels at which the wrist becomes linear. X = male, O = female. Black square = mean for males, filled circle = mean for females. The values are clearly lower for the women (from Lakie *et al.* 1984*b*).

reflex? In the medical literature, statements have long been made that resting muscle tone depends on a low-level tonic discharge to the muscles. Waller (1896) stated:

> The *musculo-tonic* action exercised by the spinal cord keeps the whole muscular system in a state of slight contraction or tone. There is no reason for attributing such action to special centres; it is in any case not to be regarded as anything beyond a slight and continuous motor discharge, and there is no reason for regarding it as different in kind or origin from the stronger and discontinuous motor discharges that cause muscular contractions.

Schmidt (1978) wrote similarly:

> Even in a relaxed limb, the motor nerves are activated at low frequency. The resulting tone is detectable as a *resistance to passive bending of the limb*.

A current textbook (Guyton 1981) states the following:

> Even when muscles are at rest, a certain amount of tautness usually remains. This residual degree of contraction in skeletal muscle is called *muscle tone*. Since skeletal muscle fibers do not contract without an actual action potential to stimulate the fibers except in certain pathological conditions, it is believed that skeletal muscle tone results entirely from nerve impulses coming from the spinal cord.

There is a similar statement in another popular textbook (Berne and Levy 1988):

> This *tone* is apparently a result of low levels of contractile activity in some motor units driven by reflex arcs from receptors in the muscles; tone is abolished by dorsal root section.

With the apparatus described above it was possible to record the EMG activities while undertaking the biomechanical measurements. With the larger forces and wide swings at or near resonance there was sometimes some activity. The phasing of this indicated that it was not a stretch reflex but a shortening reaction, which would appear to serve the role of 'taking in the slack'. It was the muscle group that was relieved of tension that became active. It could come and go

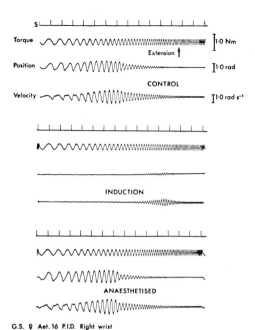

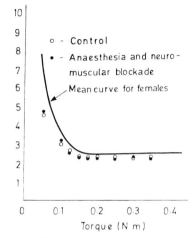

Fig. 4.15. Resonance curves of a patient undergoing anaesthesia. There is a striking increase of stiffness during induction (thiopentone and halothane), but when anaesthesia is established the record is scarcely different from that before the procedure. Other records obtained under anaesthesia have been published by Lakie *et al.* (1980a).

Fig. 4.16. Torque *vs* resonant frequency of left wrist of an anaesthetised subject. There is no change following anaesthesia and neuromuscular blockade. The figures are very close to the mean curve for female subjects. Patient, 22-year-old female, laminectomy. (From Lakie 1981, by permission.)

apparently according to the mental state of the subject and was totally absent in about half the normal subjects tested. For moderate and small movements there was silence. The method is a sensitive one: movement of one finger can cause an easily detectable discharge. At no time in thousands of observations were stretch reflexes recorded from normal relaxed subjects. Accordingly it was natural to question widely held assumptions about the basis of tone.

It was decided to proceed with a more rigorous investigation. The instrument was taken to the anaesthetic room of one of the local hospitals. Lifting an anaesthetised patient is like lifting a sack of potatoes; this is because the adaptive postural reactions are in abeyance. In an anaesthetised person all neural activity to the muscles has not ceased; with most commonly used anaesthetics, breathing continues. The patency of the upper airway is maintained in the conscious subject, irrespective of the position of the head and neck, by active mechanisms that involve muscles inserted into the hyoid and thyroid cartilages, the strap muscles. With thiopentone anaesthesia the activity changes from tonic to phasic, in step with inspiration. Airway obstruction is frequent and accompanied by significant increases in phasic muscle action which may not overcome the obstruction (Drummond 1989).

As it cannot be assumed that neural activity affecting the limbs is necessarily abolished in anaesthesia, measurements were made of wrist resonances in patients who were being subjected not only to general anaesthesia but also to full doses of neuromuscular blocking agents for their surgical operations (Fig. 4.15). An example of the relationship between torque and resonant frequency is shown in Figure 4.16. The general conclusions were that there was no reduction in tone as a result of anaesthesia. Evidently, the drugs used to give 'relaxation' are successful because they interfere with reflexes which would otherwise occur during operative procedures. By abolishing these reflexes the surgeon can obtain access without muscular responses which otherwise would occur. The common usage, 'relaxants', is misleading.

Such widely held views about the nature of resting muscle tone, quoted above, deserve detailed consideration.

1. The principal source of the opinion was an experiment by Brondgeest (1860). The quotation is from Luciani's account (1915):

> If the lumbar plexus of a frog is cut on one side, after its spinal cord has been divided higher up so as to paralyse voluntary movements, and the animal is suspended vertically by its head, the two hind limbs of the animal take up essentially different positions. The leg on the side on which the nerves were cut hangs fully extended, *i.e.* the muscles are flaccid, whilst that of the other side, on which the nerves are intact, is slightly flexed owing to the tone of the muscles.

It is indefensible to draw conclusions about physiological processes in man from such dated and indeed irrelevant data.

2. The Sherringtonian school in Oxford discovered the stretch reflex. This was a most notable and important basic advance. However, the general position was reductionist: *i.e.* having discovered one significant factor, it was held to be the basis of all tone. Thus, in the work of that group (Creed *et al.* 1932), the term 'tone' in the index is suffixed '(stretch reflex)'. Because of the tremendous prestige and influence of these workers their view has been widely accepted.

3. Since the introduction of needle-type EMG techniques, however, a different view has been current in the circles of clinical electromyographers. Clemmesen (1951) wrote:

> The normal resting muscle presents no potentials whatever, and no evidence at all of any slight sustained stimulation. 'Motor tone' at rest does not exist, passive elastic forces only are the causes of the tensions of the resting muscle.

Ralston and Libet (1953) reported on numerous studies in normal subjects and amputees as follows:

> It has been uniformly observed that, in the relaxed human subject, sitting or lying down, there is no detectable persistent background of electrical activity in the many muscles that have been examined in the trunk, limbs and jaws. These muscles are electrically silent unless the individual tenses them.

Basmajian (1957), who has undertaken EMG studies using needles in many normal people, reported:

> In no normal muscle at complete rest has there been any sign of neuromuscular activity, even with multiple electrodes.

These views are entirely consistent with the data presented above. They have considerable implications for the numerous and often repeated claims of the efficacy of relaxation therapy and biofeedback. If anaesthesia with neuromuscular blockade does not cause any reduction of tone below that of a normal resting subject, how can psychological suggestion be effective?

There are numerous small publications with titles such as *Teach Yourself to Relax*, and newspapers often print articles on the same theme. One quotation, from a book on yoga (Marshall 1975), may suffice:

> Most people find complete relaxation a difficult thing to achieve. You can't force yourself to relax no matter how hard you try—it just doesn't work. You have to *allow* yourself to relax by letting go of every single muscle in the body and the face, and then clearing the mind . . . The ability to relax completely at will, both physically and mentally, is something that only few people can do, but as you will see this can be achieved easily once you know how. It simply entails lying down on the floor for a few minutes in the Corpse posture.

What can these procedures achieve? The answer may well be that in people who are 'of a nervous disposition'[4] there is difficulty reaching the basal level so naturally achieved by others.

The author studied 'deep relaxation' in 10 people. In these sessions, run by an experienced therapist, the subjects were asked to lie on a mattress, covered with blankets, and told to close their eyes and listen to soothing words on a tape recorder. The tremor level was recorded using an accelerometer and an instrument which integrated the output over 5s. Measurements were made of the tremor of an outstretched finger before and after the relaxation session. One man, the headmaster of a tough inner-city school who had just come from a staff meeting, was 'shaking like a leaf' beforehand. Half an hour later his values had fallen dramatically. In the others there was no change.

The shortening reactions described above may be brought up, if absent, by other neuromuscular activity. It has long been known that with voluntary activity in one area there may be recruitment of muscle activity in distant parts of the body. Some of these effects were elegantly studied by Benson and Gedye (1961)—their arrangements are shown in Figure 4.17. Perhaps relaxation therapy might be useful, if such recruitment is excessive, but the author is not aware of any relevant neurophysiological investigations.

Tauber *et al.* (1977) studied the EMG during sleep. They recorded from 17 postural muscles and found no activity. However, in the mentalis of the face there was tonic activity during most periods of sleep when eye movements were absent. This was reduced during the periods of rapid eye movements.

---

[4]French (1969) quoted what Adair wrote in 1786: 'Upwards of thirty years ago, a treatise on nervous diseases was published by my quondam, learned and ingenious perceptor, Dr Whytt, Professor of Physick, at Edinburgh. Before the publication of this book, people of fashion had not the least idea that they had nerves; but a fashionable apothecary of my acquaintance, having cast his eye over the book, and having often been puzzled by enquiries of his patients concerning the nature and causes of their complaints, derived from thence a hint, by which he cut the Gordian knot—"Madam, you are nervous"; the solution was quite satisfactory, the term became fashionable, and spleen, vapours and hyp, were forgotten'.

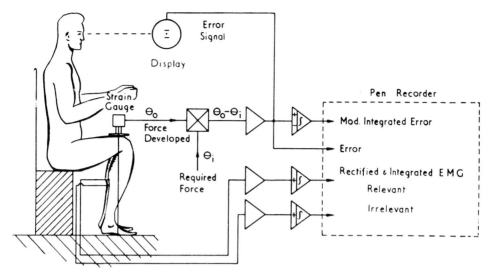

Fig. 4.17. Experiment to measure 'irrelevant' muscle activity when engaged in a task. The subject presses down with his right foot and the resultant tension deforms a spring bar placed horizontally across the lower part of the thigh. The person is required to reach a certain level of tension and an error signal is fed back to him. The EMG activity of the opposite calf, and sometimes of other sites, is recorded. The work was undertaken for the RAF. There was an interest in pilots who during difficult manoeuvres became so tense that their ability to control the aircraft was impaired in them; there was greater than normal contralateral activity during the task. (From Benson and Gedye 1961, by permission.)

Damping

At resonance the inertial and elastic forces are in balance and cancel one another, and damping alone determines the motion for a given rhythmic torque. With zero damping the amplitude and velocity would be infinite; the damping is inversely related to the sharpness of tuning which may be expressed by the quality factor Q[5]. This parameter can be obtained from resonance curves by dividing the resonant frequency by the difference in frequency between the two points where the peak amplitude is reduced to 70 per cent of that at resonance.

Another method of obtaining Q is shown in equation 12 (p. 75). At the wrist, Q was commonly found to have a value of about 2, corresponding to a damping ratio of about 0.25 (Burton 1968). For critical damping, at which level no resonance would occur, the damping present would have to be increased fourfold. The wrist is thus significantly underdamped. Values for Q at the elbow are given in Table 6.I. They are of the same order of magnitude.

Damping may be investigated by plotting the peak velocity at resonance against peak torque. With purely viscous damping the points would lie on a straight line; this is approximately true (Fig. 4.18). The velocities attained by women for a

[5]Also known as the 'magnification factor' or 'dynamic magnifier'. It is the ratio of the vibrational amplitude at resonance to the static displacement that would be produced by the steady application of a force equal to the peak value of that applied rhythmically. Such a method of measurement is perhaps not very suitable for investigating the postural system because of thixotropic effects. Static determinations can thus be erroneous.

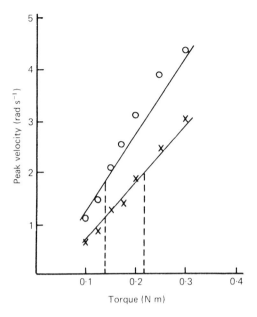

Fig. 4.18. Peak velocity at resonance of 12 male (X) and 11 female (O) subjects—left wrists. The mean values fall on an approximately straight line. The critical peak torques at which the resonant frequencies became constant are indicated by the dotted lines. In the two instances this occurred at about the same peak velocity (1.8 rad s^{-1}) (from Lakie *et al.* 1984*b*).

given force are higher than those for men, again no doubt because they are less muscular.

Viscous damping is unlikely to arise in relaxed muscle. Hill (1968), working with frog muscle, found resistance corresponding to simple friction rather than viscosity. There is almost no friction in a healthy joint. The sliding friction of tendons moving in their sheaths may be quite low but the area of the surfaces is large; motion between fascial planes too is likely to be significant.

Summary

The procedures described in this chapter enable damping to be measured while stiffness can be estimated although not measured in absolute units. Stiffness comparisons, based on measurements of the square of resonant frequencies, can be useful for the statistical comparison of different populations and for following changes in any given person. For many situations such comparisons are adequate. With the developments described in Chapter 6 it is possible, where necessary, to obtain absolute values.

5
FEEDBACK-INDUCED CHANGES

Negative feedback occurs when a disturbance gives rise to a control which causes a reaction opposing the disturbance (Fig. 5.1). In the human body there are many negative feedback systems. One well known example is the regulation of blood pressure by the carotid sinus mechanism; a rise of pressure in the artery increases the discharge through the nerve to the brainstem and this reflexly causes dilatation of the arterioles.

Normal movements can be executed smoothly and postures are maintained without oscillation. The stability evidently depends on internal forces in muscle and, it is thought, on negative feedback. The situation can be changed by subjecting a limb to a source of force that is controlled according to the motion.

Negative velocity feedback
If a velocity signal is fed back to the amplifier supplying the motor so that a force is generated opposing any movement of the motor shaft, this is negative velocity feedback. There is effectively increased viscosity and stability. If a lever attached to such a system is moved by hand, the impression is that of stirring treacle. The first time this was tried the velocity signal was obtained by differentiating the position signal from a somewhat imperfect potentiometer. The sensation on moving the lever was that of stirring gritty porridge with a 'spurtle' or potstick.

Positive velocity feedback
If the connections are reversed, force is added according to the velocity attained as the hand is moved; there is positive velocity feedback. If the person then attempts

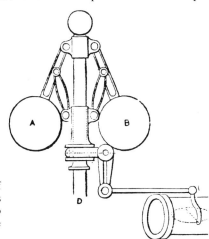

Fig. 5.1. Steam engine governor, an early form of negative feedback. As the shaft 'D' rotates, the weights 'A' and 'B' spin round and move outwards due to centrifugal force. As they rise they progressively shut the valve supplying steam to the engine (from Evers 1880).

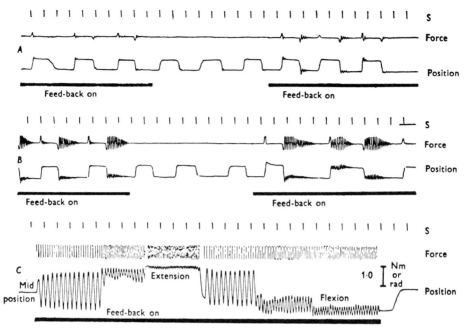

Fig. 5.2. Subject makes abrupt flexion and extension movements with and without closure of a feedback loop in which a velocity signal is sent to the motor causing a torque to be applied assisting the motion. *Top:* at a gain of 0.05Nm / rad s $^{-1}$ there is some increased overshoot. *Centre:* feedback gain is doubled—pronounced ringing at 9Hz. *Bottom:* further increase of gain—continuous oscillation at 3 to 7.5Hz. Force is delivered as square waves because the amplifier saturates. Oscillations are slowest with wrist held naturally, increased rates are seen when it is extended or flexed (from Walsh 1970*a*).

to move his hand quickly, accurate motion becomes impossible. There is overshoot and a variable number of oscillations at the end of the movement (Fig. 5.2).

With high gain in the feedback loop the motor and hand may start to oscillate spontaneously; the man/machine combination forms an oscillator. With lesser gain the oscillations may not start spontaneously but become self-sustaining after having been started by a tap. The electrical circuits saturate, and accordingly the currents through the motor alternate in a rectangular fashion. An electrical analogy is that of a circuit arranged to have negative resistance. The torque generated by the motor causes the hand to move until its momentum is restrained by stretching the tissues and a rebound occurs. As soon as the velocity reverses, the current in the motor (and hence the torque) reverses, pushing the hand backwards until it is again checked by elasticity when the current again reverses, and so on as long as the system is energised.

The method depends on there being some bounce at the extremity of the movement which may be due merely to the passive underdamped nature of the limb resonance or a result of a muscle reflex. There are many examples of systems in which feedback causes oscillations: one well known instance is a puppy trying to catch his own tail.

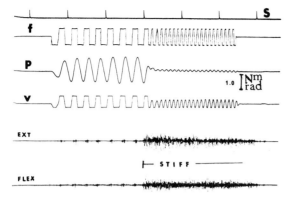

Fig. 5.3. Positive velocity feedback. The oscillations start spontaneously when the circuit is completed and are more rapid when the person voluntarily stiffens his wrist. When the patient is relaxed, the EMG channels are relatively silent. When s/he is stiff they are quite active. Downwards movement of the position trace = extension. Male, aged 20.

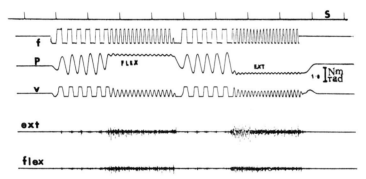

Fig. 5.4. Positive velocity feedback. With the wrist in its natural position the movements are largely dependent on the passive resonance of the tissues as the EMG channels are almost silent. When there is flexion or extension the rate of oscillation increases signifying increased stiffness and both EMG channels show activity. Male, aged 20.

With movements at the resonant frequency, there is no need to use a chirp to obtain measurements. Data can thus be obtained almost instantaneously. As the system is constantly held on resonance, and tracks it, the velocity attained is limited only by the damping. The arrangement can thus give almost immediate information about stiffness and damping and the changes which occur under different operating conditions; furthermore the electronic circuits needed are relatively simple. The square wave feedback to the motor is equivalent to a sine wave 1.27 times its size. It contains odd-numbered harmonics, but as the most significant of these is a third of the amplitutde and three times the frequency of the fundamental, the effects are negligible.

As the man/machine combination forms an electro-mechanical oscillator, held automatically on resonance, the damping coefficient (c) may be obtained by dividing the velocity by the torque. As the torque value swings through zero,

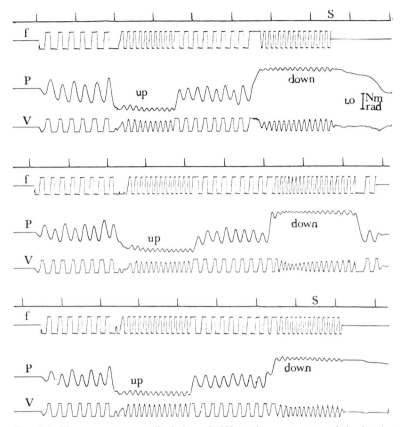

Fig. 5.5. The apparatus was tilted through 90° so the movements of the hand were in the *vertical* direction. Again the motion was slowest when the wrist was held naturally and the oscillations accelerated with flexion or extension. Male, aged 19.

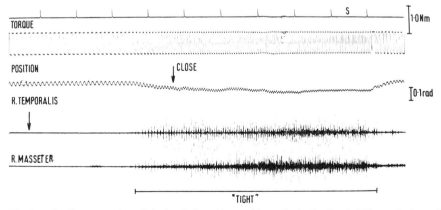

Fig. 5.6. Oscillatory motion of the jaw induced by positive velocity feedback. When asked to stiffen the person partially closes, the oscillations increase in rate and considerable EMG activity results. Male, aged 26.

however, it is usually more convenient to calculate the reciprocal of damping, the 'mobility'.

With an arrangement to use the system at the elbow, Mr Fraser McGlynn of St Andrew's University has measured changes which occur after muscular exercise. It is known that sore muscles do not usually arise from concentric contractions, where the muscle shortens, but from eccentric contractions, where the muscle is activated and is pulled out by a greater force than it is developing itself. Fewer motor units are then active and muscle biopsy studies have shown small areas of damage revealed by the use of the electron microscope. If a small sample of the muscle is biopsied a few days after the exercise, fibre necrosis and inflammation is likely to be found.

The problem comes not from lifting weights but from putting them down. Sore legs are the result of going downhill, not uphill. With eccentric exercise, the changes seem to show that the principal alteration is an increase of damping. Stiffness also increases but less strikingly. The changes take a day or two to develop fully.

When the normal arm is stiffened, the rate of the oscillations increases and there is ongoing EMG activity in the flexors and extensors (Figs 5.3, 5.4).

If the wrist is flexed or extended from its mid-position the oscillations still occur but the frequency is elevated (Fig. 5.2c). Most of the observations have been made with the movements in the horizontal plane, but the system can operate equally well vertically, and here too deviation from the mid-position is associated with an increase of rate (Fig. 5.5).

The system has been used at other parts of the body. The next chapter deals with observations at the elbow; it has been used also at the ankle and knee. With the fingers a rapid rate of oscillation is induced. The bounce in the jaw when a force abruptly reverses is small (see Fig. 9.11), but here too the system worked without difficulty (Fig. 5.6).

The non-linear properties of the limb can be explored by arranging the waveform to the motor to be varied in strength (Fig. 5.7). Here, as with chirps, the resonant frequency is seen to rise when the motion is small (Fig. 5.8).

Coupled movements of two limb segments
The elbow and wrist can be considered as two pendula in series. It is known in mechanics that coupled pendula have two modes of free vibration. The second pendulum acts as a vibration absorber when acted on by a force at its resonant frequency and the first pendulum then moves less. This system has been used to obtain stabilisation, as in some electric shavers.

Each pendulum would have its own natural frequency, which for small oscillations would be independent of amplitude. However, when coupled in series, the combination has two natural frequencies. At the lower of these, the two move in phase. At the higher they are 180° out of phase and are moving in opposite directions. Swanson (1963) commented:

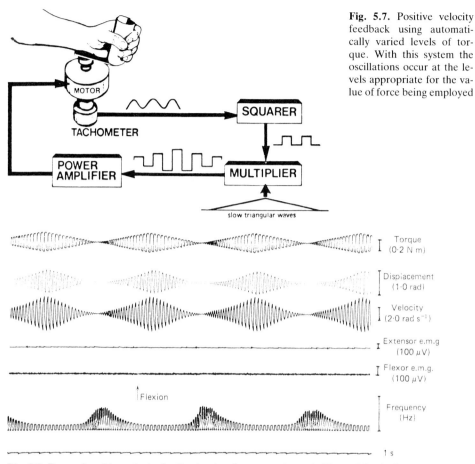

Fig. 5.7. Positive velocity feedback using automatically varied levels of torque. With this system the oscillations occur at the levels appropriate for the value of force being employed

Fig. 5.8. Ramped positive velocity feedback using the system shown in Figure 5.7. The frequency rises as the torque falls and falls as the torque rises. There is considerable asymmetry in the response—the frequency rises more rapidly than it falls, evidently due to thixotropy (see Chapter 7) (from Lakie *et al.* 1984*b*).

Fig. 5.9. Positive velocity feedback used as a therapy for a patient with a spastic left arm following a stroke. A racing car type seat was used, fixed to a metal frame to which the motor unit could be bolted to either side. The person could control the strength of the driving force by adjusting a knob on the control unit with her good hand. With this system it is essential that the arm is bandaged in position before the system is energised, otherwise the metal lever may swing round and cause an injury (from Gelman *et al.* (1975).

A human leg is two compound pendulums connected in series; the motions are nothing like so simple as those considered here, because the muscles exert moments about the pivots which vary with time in a most complicated way, but it is still true that the least effort is exerted when the leg is made to move at a frequency near a natural frequency. With an artificial leg, some or all of the muscles are absent, which makes it impracticable to force a periodic motion at any frequency other than a natural one, so, unless complicated variable dampers are fitted, the user has only one walking speed.

Usually in human movements more than one joint is concerned and the two modes of vibration can be readily observed in your own arm. With the limb dependent, flex and extend the elbow about once a second. Your hand will move in the same direction as the forearm. The segments are vibrating at the lower natural frequency. Then, shaking the limb rapidly, your hand will move in the opposite direction to the forearm. The vibration is at the upper natural frequency of the combination.

In some experiments, an alloy lever attached to a motor concentric with the elbow was furnished with a joint underneath the wrist. Forearm and hand were supported and could move independently in the horizontal plane, constituting two coupled resonant torsion pendula.

It was possible to feed back the velocity signal from the wrist in two senses. In the first, extension of the wrist led to a force from the motor extending the elbow; and with the hand flexed, a force was generated flexing the elbow. In the second system the arrangements were contrariwise. Extension of the wrist gave rise to a force flexing the elbow, and flexion of the wrist gave rise to a force extending the elbow. With both types of feedback, the system was oscillatory; and when the system was energised, oscillations started which were at the same frequency at the elbow and wrist. But the direction of the movements in elbow and wrist was either concurrent or divergent.

Either type of motion can be harmonious but, if the wrist and elbow moved at different rates, the sensations might be unpleasing. Muscle action in voluntary movements generally uses one or other of these options. To drive one segment at one rate and its neighbour at a different rate in rhythmic actions is likely to be an inefficient use of metabolic energy, tiring and fatiguing. Some muscles act over more than one joint, and their activity switches the options between the two alternatives. Through the actions of two-joint muscles the limb may behave like a pantograph.

Automated physiotherapy using positive velocity feedback
Muscular exercise has hitherto been obtainable either by voluntary action or from selected groups of muscles by the use of electric shocks. The first form depends on an effort of will, the second on a rather unpleasant procedure. Using the positive velocity feedback it is possible to exercise the limb by a method which depends neither on volition nor on electric shocks (Fig. 5.9). The system has been used in a limited number of patients with spastic hemiplegia exercising for, say, 20 minutes three times daily. The procedure is not unpleasant: indeed it is mildly pleasurable

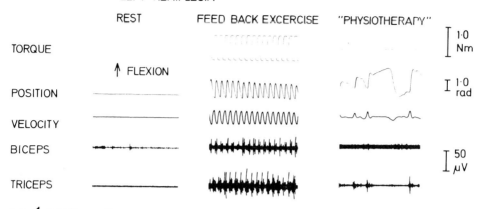

Fig. 5.10. Records from a patient with a spastic left arm as a result of hemiplegia. Positive velocity feedback evokes considerable modulated EMG activity. The passive movements that might be made by a physiotherapist produce much less. By the feedback system it is practicable for the muscles to be exercised for some hours daily.

and soporific. As it is almost silent it is compatible with holding a conversation or watching television.

Flexion and extension of the elbow in a sitting position causes substantial movement at the shoulder and rotation of the humerus. This would be undesirable if the aim was to obtain biomechanical measurements at the elbow. Chapter 6 describes a different position for such movements: but for stroke patients, who often suffer from pain in the shoulder, this motion is often regarded as beneficial and some relief of the discomfort may be obtained. The considerable shoulder girdle motion is regarded as a bonus.

The procedure certainly induces muscular activity to a much greater extent than can be provided by manual physiotherapy (Fig. 5.10). The instrument can be left running without supervision so that very large numbers of movements are feasible. Mobilisation of joints is recognised as an essential element in the physiotherapy of stroke patients to prevent contractures and changes in the joints. All of the patients treated were on a clinical plateau; in none was there any significant improvement in voluntary power, but it holds promise as a procedure for reducing spasticity.

With appropriate development the system might be of value in the training of athletes and perhaps as an aid to fitness for sedentary workers.

6
MEASUREMENTS OF STIFFNESS AND INERTIA IN ABSOLUTE UNITS

When a force acts on an object, acceleration occurs at a rate inversely related to the mass; but for angular motion, as when moving a limb, the corresponding relationship is between torque and the moment of inertia. For any part of a limb the moment of inertia is given by its mass multiplied by the square of its distance from the axis.

A human limb can be considered as consisting of many very small cubes. If the limb were to increase in size by 10 per cent in length, breadth and width, each side of each cube would increase by 10 per cent and each cube would be 10 per cent further from the axis. From such considerations it can be shown that for geometrically similar limbs, the moment of inertia will vary with the fifth power of length. If the length of a man's arm is twice that of a child, the difference in inertia will not be double, but 32-fold. Because inertia grows with us, and because most people do not have much grasp of biomechanics, its effects (although pervasive) normally go unrecognised.

Considerations of the inertia of the limbs are vital for an understanding of the physics of hand-to-hand combat. Feld *et al.* (1979) commented that:

> The techniques of karate differ markedly from those of Western methods of empty-hand combat. The karateka concentrates his blows on a small area of the target and seeks to terminate them about a centimetre inside it, without the long deliveries and follow-throughs of the punches in Western boxing. Whereas the Western boxer imparts a large amount of momentum[1] to the entire mass of his opponent, pushing him back, the karateka imparts a large amount of momentum to a small area of his opponent's body, an amount that is capable of breaking tissue and bone. A well-executed karate strike delivers several kilowatts of power over several milliseconds, quite enough to break blocks of wood and concrete.

Atha *et al.* (1985) studied the punches delivered by a heavyweight boxer, Frank Bruno, to a model head:

> Attention is drawn to the shortness of the time taken to deliver the punch from the first extension of the elbow to peak contact. If basic visual reaction time is taken as 140 ms the chances of an opponent dodging a blow he sees coming are clearly slim. His protection will instead depend primarily on the adoption of successful random evasion strategies or anticipation based on earlier, subtler cues than those associated with the start of punch delivery.

The punches delivered forces of about half a ton (5000N)!

---

[1] The angular momentum of a limb is given by the product of its moment of inertia times the angular velocity. It might have been more appropriate if the authors had referred to the kinetic energy. This figure is given by halving the product of the moment of inertia and the square of the angular velocity.

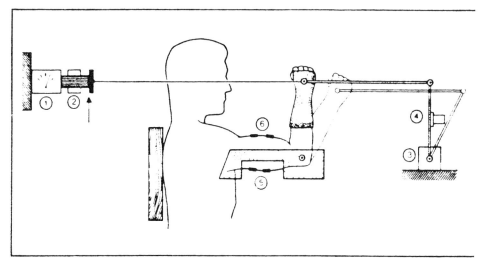

Fig. 6.1. Measurement of inertia by the quick release method. The person uses the triceps (5) to exert a certain force that is monitored by the meter (1). Quiescence of the biceps (6) is checked by EMG. Suddenly the cable s/he is pulling against is released by an electromagnetic device (2). Position is recorded by a goniometer (3), the elbow should initially be at a right angle. The motion is recorded by an accelerometer (4). The moment of inertia can be calculated by dividing the torque by the angular acceleration. This would appear to be a very sensible method provided the measurement can be made before muscle activity changes. The 'window' during which satisfactory measurements can be achieved is likely to be narrow. (From Bouisset and Pertuzon 1968, by permission.)

Previous estimates of the moment of inertia

Although the moment of inertia of a body segment is such an important parameter to be considered in any movement, there appear to have been only a limited number of attempts to obtain appropriate measurements. Braune and Fischer (1892) froze cadavers and disarticulated the members. The specimens were pierced with metal rods and suspended to oscillate about various axes. From a detailed mathematical analysis the moments of inertia were calculated:

> A normally built and strong man committed suicide. The cadaver was frozen stiff. The body segments were separated at joint levels and weighed. The distances between their centre of gravity and the adjacent joints were measured. The periods of the oscillations about axes through the centres of the joints were determined. . . The head was sawn along the strangulation furrow. The steel rod was displaced somewhat distally from where it was for the oscillations of the trunk + head. Otherwise there would not have been enough neck and the centre of gravity of the trunk alone would not have been exactly in the plane of the two axes.

A less gruesome scenario is that of the quick-release method introduced by Fenn and used for the arm by Bouisset and Pertuzon (1968). The person exerts a measured pull on a cable. When this is released the muscular tension is unrestrained and the limb accelerates. On dividing the force by the initial acceleration the required value is obtained (Fig. 6.1).

Another procedure has been to couple a limb to a stiff spring and measure the rate of the damped oscillations that take place when the system is disturbed (Peyton

1986). Muscle stiffness is regarded as negligible compared with that of the instrument. The method is interesting and simple but does not seem to have been extensively used. For wholly satisfactory operation, careful splinting or cradling of the limb would be necessary. The coupling between the limb and the instrument appears to be the main problem in view of the rapid movements imparted by the action of the spring. The system would not be suitable for investigations where there was muscular hypertonus.

Drillis *et al.* (1964), in a review, listed two other methods. The first was geometrical; its weakness was that assumptions had to be made about the density of different parts of the body. In the other system, the whole person lay on a table which was coupled on its underside to a torsion spring. The rate of damped oscillations following a disturbance varies according to the position of the limbs. Thus when the arm is moved out to extend from the axis of support, the rate is slowed. From the change of frequency the inertia can be calculated. Little use of this system appears to have been made.

Bouisset and Pertuzon (1968) made measurements of the moments of inertia of the forearms of 11 subjects and found a range of 43 to 80g m^2. Peyton (1986) studied seven subjects and obtained values from 48 to 86g m^2. I had little confidence in these surveys, because the wide range of variation between subjects (which surely was inevitable) did not come through. No consideration appears to have been given to the sex of the subjects. It was decided therefore to reinvestigate the problem using new procedures.

Measurements of hand inertia and forearm stiffness using a resonant frequency method with added inertia, or position feedback

The resonant frequency is related to inertia and also to stiffness, and the relevant expression cannot be solved by a measurement under one set of conditions alone. This is possible in principle, however, if the inertia is artificially altered and the resonant frequency again determined. The inertia of the apparatus could be increased by the use of the 'inertia bars' (see Fig. 4.5). From the resultant fall of resonant frequency the inertia of the hand can be shown by means of equation 8 (p. 49) to be given by the expression:

$$J_o = J_b \frac{f_b^2}{f_o^2 - f_b^2} \qquad \text{(equation 9)}$$

where J_o is the inertia of the hand and J_b that of the bar, the value $J_o + J_b$ being substituted for J. f_o is the resonant frequency before and f_b that after the addition. It is desirable to correct for the inertia of the apparatus, about 0.6 g m^2. Using equation 4 again, having obtained a value for inertia, the stiffness may be calculated.

An alternative procedure is to alter the stiffness artificially. The most convenient method of doing this is not to use a mechanical spring but to employ electrical feedback. If a signal that is proportional to the deviation of the hand from

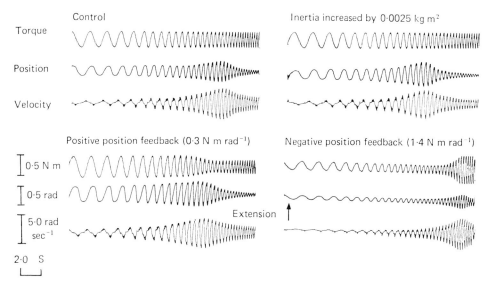

Fig. 6.2. *Top left:* no added inertia, no feedback, resonance at 2.8Hz (f_o). *Top right:* added inertia, resonance at 2.2Hz (f_b). *Bottom left:* positive position feedback, stiffness reduced, resonance at 2.4Hz (f_f). *Bottom right:* negative position feedback, stiffness increased, resonance at 4.7Hz (f_f). The calculated values for muscle stiffness using the three values are 1.23, 1.22 and 1.12.Nm rad^{-1} respectively. Considerable passive tension evidently develops in the muscles. Thus with a hand displacement of 1.0 rad the force in the stretched muscles is about 80N if it is assumed that the tendons are 1.5 cm from axis of joint (from Lakie *et al.* 1981).

its position of rest is amplified and fed back with the correct polarity to the motor, a spring is mimicked. The resultant stiffness is readily adjustable by varying the gain. The relevant expression is:

$$K_o = K_f \frac{f_o^2}{f_f^2 - f_o^2} \qquad \text{(equation 10)}$$

where K_o is muscle stiffness, K_f the feedback-induced stiffness and f_f the resonant frequency with the feedback applied. This equation too is derived from equation 8 (p. 49) and the inertia of the hand may be found by calculation.

A variant of the procedure is to use feedback in the opposite sense when the stiffness is reduced and the resonant frequency falls. In practice this is unsatisfactory because at the lower frequencies there is more likelihood of some voluntary control being exerted. There is also the possibility of significant resistance from thixotropic effects (see Chapter 7).

Examples of these types of observations are shown in Figure 6.2.

In these examples chirps have been used to determine the resonant frequencies but equally positive velocity feedback may be used with position feedback also being provided to change the effective stiffness. By varying the level of positive velocity feedback it is possible to obtain information about the system for differing degrees of movement. Using a design of this type it would be a matter of

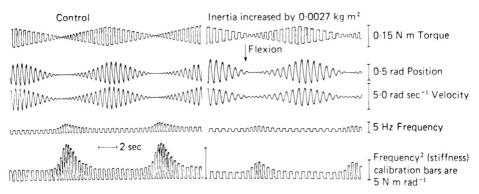

Fig. 6.3. Ramped positive velocity feedback driving the wrist at its resonant frequency. *Left:* normal. *Right:* with added inertia the oscillations are slower. The instantaneous frequency is displayed. The square of the frequency reflects stiffness, and has been calibrated in absolute units (from Lakie and Walsh 1981).

straightforward engineering and computer programming to develop an instrument that provided readouts of the stiffness, inertia and damping at different force levels.

An example of the use of ramped positive velocity feedback with and without added inertia provided by a metal bar is shown in Figure 6.3 (see also Figs 5.7, 5.8). Inertia can theoretically be increased or decreased by feeding back a signal corresponding to angular acceleration of appropriate polarity. Attempts to devise a practical system, however, were not successful.

Measurements of inertia, stiffness and damping using positive velocity and negative position feedbacks

The position of the fingers necessarily causes wide variations in the moment of inertia of the hand, for the radii of gyration of the masses alter. It was decided that observations on the forearm would produce more detailed results. As the resonance at the elbow is of low frequency it seemed appropriate to proceed by increasing the stiffness, so speeding the action. Had inertia been added the oscillations would have become even slower, with increased tendency for voluntary intervention and possible vitiation of the results by thixotropic stiffening. The technique involved the use of positive velocity feedback and electrically controlled added stiffness.

In initial experiments the person was seated and the forearm fastened to the lever, but there was inevitably rotation of the humerus as the elbow flexed and extended. Thus measurements made in this posture were confused by the consequent added inertia and the stiffness estimations were not confined to those related to the elbow flexors and extensors alone. This could be avoided by raising the upper arm to be horizontal, but the resulting postural strain might have disturbed the readings.

The person was accordingly semi-recumbent on his left side, on a mattress. The right forearm and semi-pronated hand were bandaged to an adjustable alloy

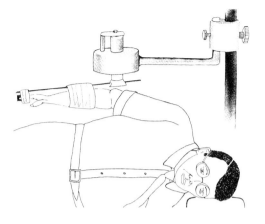

Fig. 6.4. Arrangements for studying biodynamics of the elbow. Subject lies on his side with the motor above the joint. The forearm is bandaged to an adjustable lever.

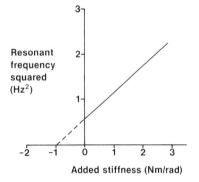

Fig. 6.5. If stiffness is added to an underdamped system the resonant frequency rises and there is a linear relationship between the square of the frequency and the addition. By extrapolating the line to zero frequency the initial stiffness can be deduced.

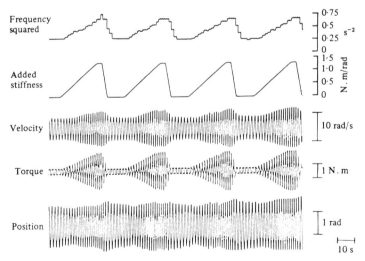

Fig. 6.6. Effect of instrumental changes of stiffness on resonant frequency. The oscillations are produced by positive velocity feedback. With a ramped increase of stiffness there is a quasi-linear increase in the square of the frequency. From the changes the absolute values of the stiffness of the upper arm and the inertia of the forearm can be deduced (from Walsh and Wright 1987a).

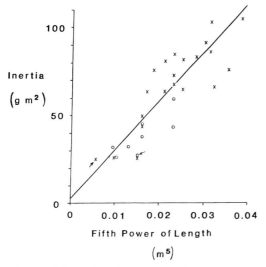

Fig. 6.7. Relationship of forearm inertia to fifth power of the length from the medial epicondyle to the tip of the extended middle finger—the 'cubit'. The correlation coefficient is 0.84. X = male, O = female. Naturally, the inertia of female arms is the lesser. The overall range of inertia values was more than fourfold. There was surprisingly only a questionable relationship between length and stiffness, r = 0.34, $p < 0.1 > 0.05$. The arrows indicate two children tested. The larger subjects had lower resonant frequencies, the relationship with the square root of the reciprocal of inertia was highly significant (r = 0.6) (from Walsh and Wright 1987a).

splint attached to a printed motor (G16M4) just above the arm (Fig. 6.4). A velocity signal from a tachometer was used to induce oscillations at the resonant frequency by the use of positive velocity feedback. The motor was also furnished with a potentiometer, and extra stiffness could be generated with negative position feedback. This was adjusted either manually or electronically with an analogue multiplier using a slowly varying ramped voltage to vary the gain.

Equation 8 can be rewritten thus:

$$K = 4 \pi^2 f^2 J \qquad \text{(equation 11)}$$

With added stiffness the resonant frequency rises and the formula predicts a linear relationship between stiffness and the square of the frequency. The initial stiffness can be determined from the data (Fig. 6.5). An example of the records obtained is shown in Figure 6.6.

The cubit, the length of the arm from the medial epicondyle to the tip of the extended middle finger, was measured together with certain other body parameters. The relationship between the length of the body segment and inertia is shown in Figure 6.7. The geometric similarity of the arms of a varied population will only be approximate, because some people with long arms may be slender, while others with short arms may be chubby. Nevertheless, it will be seen that the fifth power relationship predicted above on theoretical grounds is reasonably accurate.

TABLE 6.I

Parameters in relaxed adults (Walsh and Wright 1987a)

	Age (yrs)	Body mass (kg)	Height (m)	Cubit (m)	Forearm volume (l)	Inertia (g.m²)	Stiffness (Nm/rad)	Resonant frequency (Hz)	Q	Mobility (rad.s⁻¹.N⁻¹.m⁻¹)
Men (N 18)	41 ± 16	79.6 ± 15	1.78 ± 0.06	0.47 ± 0.02	1.61 ± 0.26	74.7 ± 16	1.03 ± 0.3	0.59 ± 0.07	1.8 ± 0.54	6.2 ± 16
Women (N 9)	37 ± 17	60.2 ± 21.6	1.66 ± 0.08	0.43 ± 0.03	1.22 ± 0.17	36.2 ± 11	0.61 ± 0.14	0.67 ± 0.12	1.2 ± 0.53	8.1 ± 2.7
Difference of means	4	19.4	0.12	0.04	0.39	38.5	0.42	0.08	0.6	1.9
p	>0.1	0.02	<0.001	<0.001	<0.001	<0.001	<0.001	>0.1	0.01	0.05
Significance of difference	Not significant	Significant	Highly significant	Highly significant	Highly significant	Highly significant	Highly significant	Not significant	Significant	Just significant

Data given as mean ± SD.

As the limb was always held on resonance the velocity was constrained only by the damping and this too could be measured in absolute units. For practical reasons an electronic circuit was used to calculate not the damping coefficient (c), but its reciprocal—the mobility.

The quality factor, Q (see p. 57), could be obtained by calculations based on the figures obtained using the formula:

$$Q = \frac{1}{c} \sqrt{KJ} \qquad \text{(equation 12)}$$

It follows that when a limb is coupled to a spring or a mass it becomes more resonant; this is one reason why one becomes more unstable than normal when carrying a load.

Data obtained by the use of these methods in adults who relaxed satisfactorily are given in Table 6.I. There are significant sex differences. There is a small but significant difference in the length of male and female forearms, and the inertia of the men is fully double that of the women. The stiffness is also substantially greater in the men, and the limbs of the men are somewhat more resonant.

Muscle strength

Muscle strength has often been measured, but much depends on the motivation of the people to get maximum contractions and on the time that the contraction needs to be held. Newman *et al.* (1984) studied the hand-grip strength of children aged five to 18:

> Human beings have long been and continue to be fascinated by their own muscular strength; they seek in particular to discover its attainable limits by all manner of weight lifting, athletic and endurance feats. Fair grounds have traditionally catered to this fascination by some measurement device where the strong and not so strong may publicly demonstrate their prowess in raising a strength indicator by the force of a blow. . . For clinical use it would seem that the average hand grip strength based on four alternate hand measurements per subject, classified by age, gives a sufficiently consistent result for most purposes. This shows an approximately linear increase through all age groups in boys. Girls record lower hand grip values than boys at all ages and these values also follow an approximately linear progression with age until the thirteenth year when they level out. The discrepancy in hand grip strength between the sexes widens thereafter throughout the teenage years.

In a study of the post-infection fatigue syndrome, muscle strength was found to be normal (Lloyd *et al.* 1988):

> An important consideration in any test of maximal voluntary strength is the degree to which the subjects recruit motoneurons of the relevant muscles. There is considerable evidence that well motivated subjects provided that they are free of muscle, joint, or other pain, are capable of sustaining voluntarily, the maximal force possible from a muscle group (that is show no 'central fatigue'). This is based upon the failure of interpolated electrical stimuli to the nerve or muscle to increase the force output from the voluntarily contracting muscle. The ability to activate a muscle maximally by voluntary effort has been documented for limb and even respiratory muscles, and has recently been shown for the elbow flexors.

Scale

There is extensive literature on allometry, the effects of body size in biology

Fig. 6.8. The figure looks convincing until it is realised that if the Lilliputians were real people Gulliver would be so overscaled that his muscles would not be strong enough to get him into the standing position. There would have been no need for them to attack him (from Swift 1838).

(Schmidt-Nielsen 1984). It is generally true that larger creatures are more heavily built and limb girth increases more rapidly than length. Thompson (1942) made famous studies of these relationships:

> It was Galileo who, wellnigh three hundred years ago, had first laid down this general principle of similitude; and he did so with the utmost possible clearness, and with a great wealth of illustration drawn from structures living and dead. He said that if we tried building ships, palaces or temples of enormous size, yards, beams and bolts would cease to hold together; nor can Nature grow a tree nor construct an animal beyond a certain size, while retaining the proportions and employing the materials which suffice in the case of a smaller structure. The thing will fall to pieces of its own weight unless we either change its relative proportions, which will at length cause it to become clumsy, monstrous and inefficient, or else we must find new material, harder and stronger than was used before. Both processes are familiar to us in Nature and in art, and practical applications, undreamed of by Galileo, meet us at every turn in this modern age of cement and steel.

Consider 10 elastic bands joined together in series; the compliance will be 10 times that of a single band, and rotary stiffness will fall with muscle length. Consider the 10 bands joined in parallel; the compliance will be a 10th of that of a single band, and rotary stiffness will rise with muscle cross-section. Consider a lever with a spring attached at a certain distance from the pivot. The further the spring is from the pivot, the greater the stiffness around it. For the joint of a limb the stiffness will increase with the distance that the bony attachment is outwards from the joint (McMahon 1975). For geometric scaling it can thus be shown that the stiffness of a joint due to the musculature will be expected to vary with the square of limb length. Thus stiffness is not expected to increase at the same pace as inertia.

For the torque developed by a muscle, there will be just two factors: (i) the cross-section of the muscle; when active but not shortening, *i.e.* during an isometric contraction, a muscle may exert about 6kg force per square cm); and (ii) the distance of the attachment from the joint. Assuming that the muscle is active throughout, its length is immaterial. The torque generated will thus go up with the cube of limb length.

It is not easy to conceive of a tribe of men who are very large or very small. Archdeacon Paley (1846) commented:

> Throughout the universe there is a wonderful proportioning of one thing to another. The size of animals, the human animal especially, when considered with respect to other animals, or to the plants which grow around him, is such, as a regard to his convenience would have pointed out. A giant or a pigmy could not have milked goats, reaped corn, or mowed grass; we may add, could not have rode a horse, trained a vine, shorn a sheep, with the same bodily ease as we do, if at all. A pigmy would have been lost amongst rushes, or carried off by birds of prey.

The author is not aware of any extensive discussion of the effects of scale in human beings, but some large men are spoken of as being 'gangling'. It is common knowledge that adolescents often 'outgrow their strength'. This may be the result of rapid and massive increases of inertia due to lengthening of the limbs without a corresponding increase in muscle mass. Adults may adjust to this situation by using lower levels of limb angular accelerations. A disproportion between limb inertia and muscle development may set a limit to the biological survival value in primitive communities of really large human beings, whilst small individuals would be handicapped by such things as the weakness of the blow their limbs could deliver (Fig. 6.8).

7
THIXOTROPY: A TIME-DEPENDENT STIFFNESS

Anyone who has made even a casual study of posture will have become aware of a certain 'cussedness' about the subject. The limb of a decerebrate cat, on being displaced, may return to its original position or remain in the new place. Classic neurophysiological studies, undertaken mostly before the era of electrical recording, spoke of the length of a muscle as being defined by the stretch reflex; but at the same time this had to be reconciled with accounts of lengthening and shortening reactions in which a limb would remain in the new position after being moved.

Earlier work has sometimes alluded to plastic effects, and Mosso's comparison of his results with the behaviour of butter (p. 29) cannot be explained if the only properties possessed by the postural system are stiffness, viscosity and inertia.

Non-linear systems are more difficult to analyse than linear ones but they are all but universal. Galileo (1564–1642) observed that the swing of the chandeliers in the cathedral at Pisa took the same time for small and big oscillations. For a pendulum this is only an approximation and Galileo's timing depended on feeling his own pulse! As even the simple pendulum has a rate which varies somewhat with the size of the swing, it is only expediency or laziness which treats the oscillations as isochronous.

Stiffening with small forces
It has been seen that the relaxed wrist becomes stiffer if the applied rhythmic force drops below a certain level (pp. 50–54). By making observations on anaesthetised patients it was shown that neural activity was not responsible for this effect, but the source of the phenomenon was obscure. The apparatus could provide not only the rhythmic force but also an extensor or flexor bias so that the swings during a chirp took place around a changed mid-position. It was then found that the stiffening still took place (Fig. 7.1). This indicated that the effect was a property of the tissues and was not due to any special anatomical factors related to one particular angle at which the system was being operated.

Loosening caused by a disturbance
The situation was clarified by an experiment in which the wrist was subjected to a rhythmic torque, 'driving force', of fixed frequency, 2.5 to 3.0Hz, at an amplitude level at which the resonant frequency had previously been found to rise. Arrangements had been made to interpolate one or more cycles of a stronger force. Results of one such experiment are shown in Figure 7.2. The 'stirring force' was phase-locked to, and took the place of, the sinusoidal torque presented before and

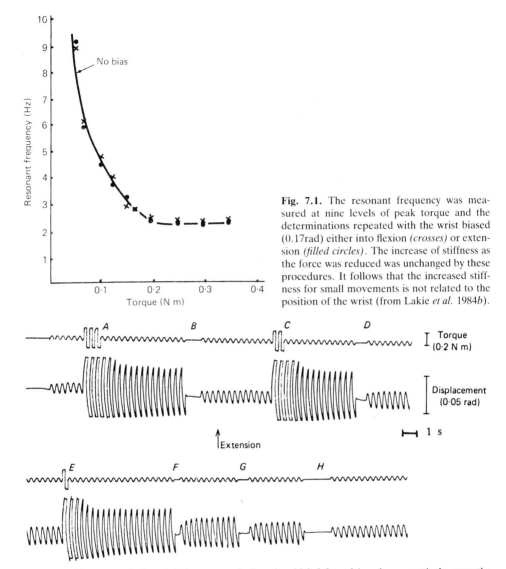

Fig. 7.1. The resonant frequency was measured at nine levels of peak torque and the determinations repeated with the wrist biased (0.17rad) either into flexion *(crosses)* or extension *(filled circles)*. The increase of stiffness as the force was reduced was unchanged by these procedures. It follows that the increased stiffness for small movements is not related to the position of the wrist (from Lakie *et al.* 1984*b*).

Fig. 7.2. Thixotropy. *A*, *C*, and *E*: large perturbations in which 3,2, and 1 cycles respectively cause the same increase in amplitude. *B*, *D*, *F*, *G* and *H*: driving force is cut out for 3,2,1,2 and 5 cycles respectively. In *B* and *H* there is a reduction of amplitude to original level, *D*, *F* and *G* cause a partial reduction (from Lakie *et al.* 1984*b*).

after the disturbance. It increased the motion during its period of application but, when the torque reverted to its previous value, the oscillations remained at a very clearly elevated level. The increase in amplitude persisted for as long as the system was kept in motion and was commonly associated with a feeling of a reduction of the internal friction of the forearm as though the limb had been 'lubricated with sewing-machine oil'. If the force was removed for 0.5s the motion stopped completely; when the torque was reapplied the motion was reduced a little, but it was still clearly greater than in the initial period. If the motor was switched off for 2s or more, however, the amplitude reverted to about its initial level; the system had stiffened. It is clear that there is a memory for past events, but that this fades rapidly.

The interpolated waveform has normally been a rectangular wave but can equally well be sinusoidal or triangular provided it is significantly more powerful than the basic rhythmic torque. Square waves have been preferred merely because in the recordings the interpolation is somewhat clearer.

The disturbances did not need to be passive, for voluntary movements were equally effective. After a voluntary movement the hand rarely came to rest exactly in its initial position; there was generally a shift of baseline. The voluntary movements did not have to be large; indeed it appeared that any movement that took the system beyond its stiff region could be effective.

If the driving force was too small the stirring force gave no increase or a transient increase that was not self-perpetuating. Presumably the driving force must be adequate to maintain the system in its loosened state by the movements that it generates. If the driving torque was rather too large the oscillations of the hand steadily increased over several cycles. The system had then loosened itself spontaneously and the application of a stirring force thereafter produced no further increase of amplitude.

Neural influences?
In many observations, EMG activity recordings showed no sign of any activity corresponding to a change of state; indeed there was silence. Observations were also made on seven anaesthetised patients (Fig. 7.3). In each patient, thixotropic effects were easily generated both before and after the administration of neuromuscular blocking drugs. It is thus clear that these non-linear properties represent passive effects in the limb.

Changes induced by cooling
Athletes 'limber up' before running. Physiologists have long believed that cold increases muscle tone, and people who are cold often experience a sensation of stiffness. In a classic description of a victim of hypothermia, Laufman (1951) observed that the neck muscles were rigid and the elbows could be flexed only with great force.

Changes of muscle tone in human refrigeration
Cardiac surgery patients commonly have their body temperatures lowered; there is no shivering, as neuromuscular blocking agents are employed. For coronary bypass operations the blood temperature may be reduced to 30°C to protect the brain. For other cardiac surgery, such as operations for coarctation of the aorta, the temperature may be dropped to 15°C. The person then feels 'doughy'.

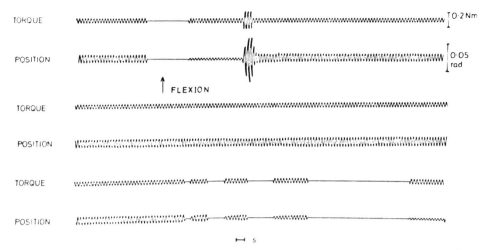

Fig. 7.3. Thixotropic changes in an anaesthetised patient after intravenous administration of 35mg suxamethonium. The increase in amplitude after the application of a stirring force persists indefinitely. The amplitude returns to its original level when the driving force is omitted for more than 2 to 3s. The tracing has been cut into three sections and represents a continuous record. Male—aged 42—laminectomy. (From Lakie 1981, by permission.)

Fay and Smith (1941) reported on a series of 42 people who were refrigerated:

> As the body temperature was lowered the muscles of the extremities assumed a waxy stiffness and when percussed contracted slowly, with subsequent slow relaxation and return to their normal contour. Spontaneous myokymia was observed in patients with temperatures as low as 84 F; if the phenomenon was not present at this temperature level, it occurred when passive motion of the joint was made. The patients assumed a characteristic posture denoting cold, with arms held close to the sides and hands against the upper portion of the chest and beneath the chin. The upper and lower extremities assumed a flexed position, and considerable effort was required to extend them. In a strong patient, it was frequently difficult to extend the arm sufficiently to take blood pressure readings. The wrists were flexed; the hands did not form the fist position but took a 'clawlike' attitude. The metacarpal phalangeal joints remained extended, though the interphalangeal joints remained flexed. The patients attempted to curl up into a ball during the shivering phase. When the rectal temperature was reduced below approximately 88 F, shivering ceased and the patients were comparatively comfortable, though not relaxed, for the increased muscle tone remained . . . In every instance retrograde amnesia existed for the period corresponding to that of refrigeration. Although memory and reasoning powers are retained during the period of refrigeration, not a single patient has been able to recall later, under normal temperatures, incidents arising while under refrigeration, when the rectal temperature was reduced to 92 F or below.

Myotonia

In the myotonic disorders there may be increased sensitivity to cold. There is a story of a window cleaner with a myotonic condition being unable to get down because the cold had provoked an attack and rendered him immobile.

Dr Thomsen in 1876 described the condition now known by his name, Thomsen's disease. He traced the disease back in his own family to 1742 and in certain cases it had shown up in infancy. For those who avoid eponyms it is known as 'myotonia congenita'. A Prussian army doctor would not accept a certificate that

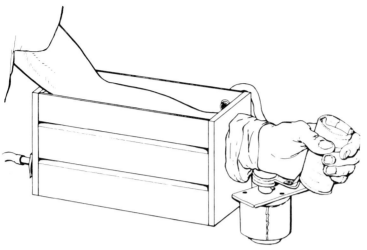

Fig. 7.4. Apparatus for cooling the forearm whilst allowing the biodynamics of the wrist to be evaluated. Water at different temperatures is pumped through the tank. A seal formed from a surgical rubber glove prevents water escaping. A small printed motor (G6M4) provides torques. (From Lakie *et al.* 1986*a*, by permission.)

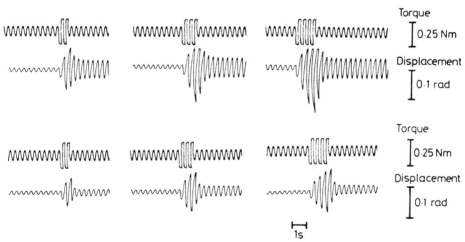

Fig. 7.5. Influence of cooling on thixotropy. The upper records are at room temperature—a large increase in looseness follows perturbations of 2, 3 or 4 cycles respectively. Cooling for 40 minutes at 8°C reduces the change that follows a perturbation but does not change the preceding stiffness. The EMG remained silent. (From Lakie *et al.* 1986*a*, by permission.)

his son suffered from this muscular defect and there are recorded instances of soldiers being punished because of myotonia which for instance made them unable to raise their hand to salute or free a finger from the trigger of a rifle (Thomasen 1948). People with this disease have difficulty in getting the muscles to relax and starting to run or walk briskly; the movements may resemble a slow-motion film. Myotonia can cause delay in opening the eyes after strong closure. Alcohol may relieve the symptoms and the sequence 'dinner, wine and dance' may thus be possible.

Dystrophia myotonica is an 'autosomal dominant' characterised by hypotonia. The resting potential across the muscle membrane is lower than normal. There is muscular atrophy and weakness. The face is characteristically expressionless and there may be bilateral drooping of the eyelids. There is difficulty in retracting the corners of the mouth and pursing the lips. As the masseter, temporalis and sternomastoid muscles are wasted the face has a haggard appearance, and in males there is frontal baldness. Many patients have a nasal and dysarthric speech, not only because of facial weakness but also because the palate, tongue and pharynx are affected. Other abnormalities are often found too. Gonadal atrophy, cataracts and mental deficiency may be found.

In paramyotonia the effects of a cold wind may be seen in a spasm of the facial muscles. All muscles may become weak and stiff when cooled. A patient with this condition was investigated experimentally by French and Kilpatrick (1957). Power may not be regained for several hours after cooling a limb. Even when reheating is by immersion in warm water it may be an hour before there is full recovery.

Experimental cooling of the forearm
To investigate the effects of local cooling in normal subjects, the forearm was placed in a specially constructed tank through which water of different temperatures could be arranged to flow. The hand, which was coupled to a small printed motor (type G6M4), protruded through a seal (Fig. 7.4). When cool water was used (*e.g.* 5° to 10°C), changes began to be observed in a few minutes. In the cooled arm, thixotropic effects were altered in that the loosening following a disturbance was greatly reduced (Fig. 7.5). At the same time the ability to make voluntary movements was impaired. Rapid oscillatory movements could no longer be made; there developed what may be called 'cold adiadochokinesia', and even the accomplishment of not especially fast movements was laboured (Fig. 7.6). Electrical stimulation was also used; the resultant twitches were slowed particularly in relaxation.

The concept of thixotropy
The stiffness of a tissue is inversely related to the extension that occurs in response to an applied force. According to Hooke's law (p. 172), the same value would be obtained no matter what force was employed. Several workers, including Hill (1968), have shown that the stiffness of a muscle is greater for small than for large stretches. The 'short-range elastic component' is found for stretches of up to 0.2 per cent of muscle length. In the noninvasive experiments on the wrist using chirps it has been noted that a similar relationship can be demonstrated in the intact human being, and the same order of magnitude is involved (pp. 51–52). Thixotropic effects have been demonstrated in isolated muscles by Lakie and Robson (1988*a*). It is suggested that when the agitation is below a certain level the system undergoes a

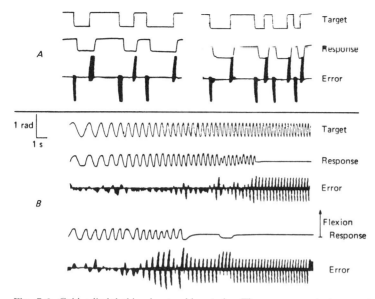

Fig. 7.6. Cold adiadokokinesia—tracking tasks. The person controls a spot on a large dual beam oscilloscope which s/he attempts to keep aligned with a second spot controlled by a waveform generator. An electronic circuit computes the difference between the two and this—the error—is displayed as a dark area. *A:* 'ballistic' movements—unpredictable low-frequency square waves were accurately tracked before *(left)* and after *(right)* cooling. *B:* the generator provides sine waves which increase from 0.3 to 3.0Hz in 15s. At room temperature these could be followed to at least 2Hz *(top)*. After cooling this ability was seriously impaired *(bottom)*. Reciprocating movements could not be generated at a frequency much greater than 0.5Hz. Cooling prolongs the relaxation phase and upsets the fine balance between the contraction time of the agonist and antagonist upon which rapid oscillating movements depend (from Lakie and Walsh 1982).

transformation from a sol to a gel state. With larger rhythmic torques the system becomes a sol and is consequently looser.

Stiffness that depends on the past history of movement characterises thixotropy (from the Greek θίξις, 'touch', and τροπή, 'turning').

The term was introduced by Péterfi (1927), who noticed changes on stirring, with a needle under the microscope, the cytoplasm of sea-urchin eggs. The gel turned to a sol but reverted to being a gel after a period when it was not disturbed. This is a common property of large molecules in solution. Agitation disrupts bonds which form between the molecules and which take time to reform once the motion ceases. This behaviour is found in materials such as clays, paints and sauces. For tomato ketchup there is the ditty 'it's shake and shake the ketchup bottle, none will come and then a lot'll'. Sometimes thixotropic compounds are used in pharmaceutical preparations. Bentonite is added to calamine lotion to improve its physical properties, so that it does not run off the skin. The chemist's instructions to 'shake the bottle' must be taken seriously. Blood behaves thixotropically: on standing, red cells aggregate to form rouleaux. Changes of this type have been described in mucus. Gastropods move using a single appendage, the foot. Observations of a

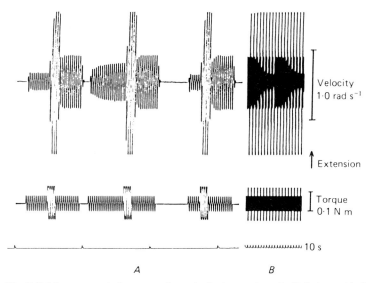

Fig. 7.7. Measurement of memory time. An instrument was built that provided a programmed series of rhythmic torques. Each cycle provided small torques, then 1 to 4 cyles of a larger force, then the small torques again and finally an interval when no force was applied; another cycle then started. A: intervals were set manually. B: the intervals increased automatically, slower paper speed. From the pattern the memory time could be measured; the shrinkage of the inner dark area indicates the increasing stiffness as the interval lengthens progressively (from Lakie *et al.* 1984a).

certain slug have shown that its mucus can change from being a glue, assisting adhesion, to being a liquid and so allowing for progression. The alterations are caused by a muscular wave sweeping forward on the foot (Denny 1980).

Hill (1968) attributed the short-range stiffness he observed to the formation of cross-bridges between the actin and myosin filaments of the muscles. This may be the basis of the thixotropic changes. It may be supposed that motion beyond a certain range tears the cross-bridges, which take time to reform once the agitation ceases.

Measurement of memory time
An instrument was built to supply a programmed series of rhythmic torques. The cycle consists of a series of small sinusoidal torques, then one to four cycles of a larger force, then the small torques again, and finally a period when no force is applied before the chain of events is repeated. The interval may be preset. Alternatively it may increase automatically from 0 to 2s in incremental steps of 0.25s, with one increment each time the system completes a cycle until the circuits are reset and the whole sequence is restarted. The stiffening following the different intervals gives a measure of the time course of the thixotropic memory (Fig. 7.7).

Biodynamics of the human hip
If indeed the phenomena described depend on a general property of skeletal

Fig. 7.8 *(below)*. The leg is supported from the ceiling by cords so that it can swing in the horizontal plane. A printed motor is positioned over the hip joint and drives the leg by means of a lever furnished with a yoke. For the measurements of free swings an induction generator, sensitive to acceleration, takes the place of the motor (from Walsh and Wright 1988).

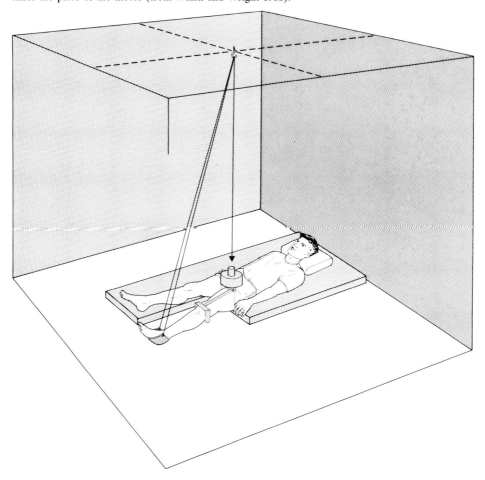

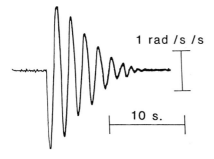

Fig. 7.9 *(left)*. The leg is manually displaced and then, on release, it swings for a number of oscillations. At first the rate is about 0.5Hz but when the oscillations become small they are faster, an effect perhaps due to thixotropic stiffening. The angular accelerometer can detect oscillations which are too small to be easily visible. The oscillations fall away progressively but commonly last for about 10s. This procedure, which requires only simple instrumentation, might be worth exploring as an alternative to the Wartenberg test—and might be especially interesting in cerebral-palsied children with spasticity of the adductors of the hip (from Walsh and Wright 1987c).

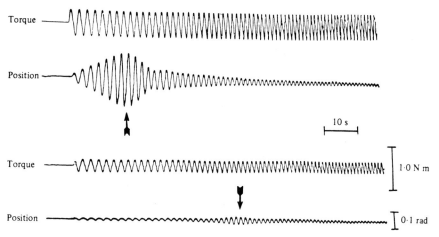

Fig. 7.10. Chirps at two force levels. The resonant frequency with the higher force was 0.47Hz, when the force was halved it rose to 0.73Hz. Arrows indicate resonance. Abduction downwards. Male, aged 33 years (from Walsh and Wright 1988).

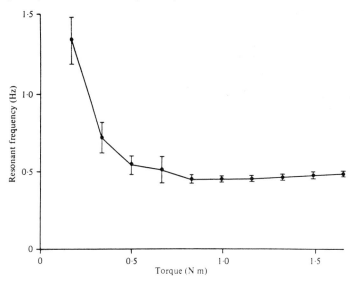

Fig. 7.11. Resonant frequency *vs* peak torque for six subjects. The values were essentially constant at the higher torque levels but climbed as the torque was reduced indicating stiffening (from Walsh and Wright 1988).

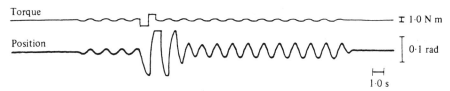

Fig. 7.12. Loosening of the system by one cycle of an interjected square wave. There was a rest of several seconds before the record was taken. Abduction upwards (from Walsh and Wright 1988).

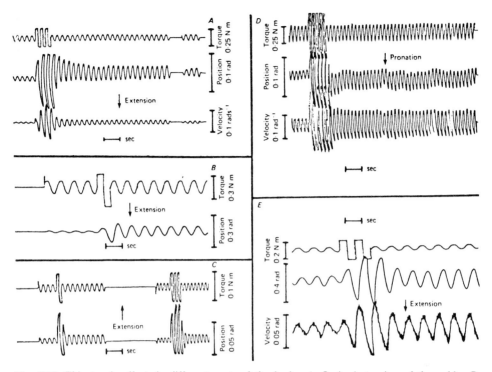

Fig. 7.13. Thixotropic effects in different parts of the body. *A:* flexion/extension of the ankle. *B:* flexion/extension of the knee. *C:* flexion/extension of the proximal interphalangeal joint of the index finger. *D:* pronation/supination of the forearm. *E:* flexion/extension of the right elbow (from Lakie *et al.* 1980*b*).

muscle they should be demonstrable elsewhere than at the wrist. A detailed study of the biodynamics of the hip was undertaken as the hip is a large proximal joint, the anatomical arrangements of which are very different from the wrist. The apparatus is shown in Figure 7.8.

In such a situation the hip is very clearly an underdamped system. Following a manual displacement, a train of decrementing oscillations may be observed (Fig. 7.9). Anyone who doubts that relaxed muscle has stiffness should feel the resistance encountered on manually abducting the hip with this type of arrangement.

A resonance was readily recordable with the use of rhythmic torques, and as with the wrist there was evidence of substantial non-linearity. An example of chirps at two force levels is shown in Figure 7.10. The resonant frequency was essentially constant at the moderate and higher levels of torque but rose for the smaller values (Fig. 7.11). When in addition to the rhythmic torque the motor also supplied a steady bias, so that the swings occurred around a different central position, the stiffening (as shown by the elevation of resonant frequency for the smaller forces) still occurred.

These results were very similar in their general features to the behaviour of the wrist, and a search was made for evidence of thixotropic behaviour. At appropriate force levels and with suitable adjustment of a driving force it was found that stirring with a perturbation loosened the system, as with the wrist (Fig. 7.12). Similar recordings were obtained in two patients with artificial hip joints.

Observations in association with Dr P.-M. Gagey of Paris were also made of the biodynamics of the hip joint using axial rotation of the leg. A footplate attached to a G16M4 motor allowed resonances to be recorded. For rotations of moderate degree the resonant frequency was in the range of 3.5 to 4.5Hz. The inertia of the leg for axial rotation is naturally much less than when the limb is abducted and adducted or flexed and extended.

Thixotropy at other sites

Thixotropic effects similar to those seen at the wrist and hip have been observed with modified apparatus in other parts of the body (Fig. 7.13).

The tendon of the biceps femoris may readily be palpated through the skin at the back of the knee. Using a small printed motor with a shaped lever, low-frequency (3Hz) vibrations were applied to push in the tendon rhythmically. Thus the muscle could be stretched when the rest of the limb was stationary. If the knee was then moved to and fro once or twice, the stretching was greater. When the knee was returned to the original position, however, the motion of the lever was clearly greater. The loosening had occurred as in the experiments already described.

Lakie and Robson (1988b) performed a similar experiment with an instrument that delivered small controlled taps to the extensor digitorum communis muscle:

> A steady biasing torque exerted by the motor pushed against the muscle and kept the lever in place. The finger could be stirred in the usual way. When taps were superimposed on the steady torque a transverse oscillation ensued; the resonant frequency of these transients increased as the rest time increased. The frequency could also be elevated at will by voluntary contractions of the muscle. Under these circumstances the joints and tendons could not be involved in the stiffening process.

However, Wiegner (1987) has observed thixotropic effects in the isolated ankle joint of the rat. Synovial fluid is known to have non-Newtonian properties. Its complex viscoelastic properties, evidently dependent on the hyaluronic acid content, have been documented by Thurston and Greiling (1978). Joints may make a contribution to thixotropic effects but this may be relatively small. Lakie and Robson (1988b) searched for thixotropic effects in the terminal joint of the middle finger in a position in which the muscles were inoperative (pp. 103–104), and found none:

> There was thus observable thixotropy only when the musculature was involved; it was not seen when the skin, joint and tendon alone were tested.

Thixotropy in invertebrates

Observations were also made on amputated legs from edible crabs. Printed motors are not manufactured in suitably small sizes, but basket-wound motors can also be

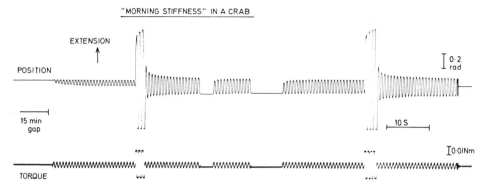

Fig. 7.14. A surplus refrigerator was fitted with a glass window and observations were made on the amputated legs of crabs at about 5°C. Thixotropic effects were seen but the memory time appeared to be about 15 minutes.

used to provide rhythmic torques (Fig. 7.14). For a high degree of stiffening to follow large movements, a 15-minute rest period may be needed.

Some observations were also made on scallops. Here too thixotropic effects could be demonstrated.

It has been known for a considerable time that some invertebrate muscles have 'catch' mechanisms, and shortening may persist for long periods after a contraction. It would be interesting to know the relationship of this phenomenon to thixotropy.

Biodynamics of the eye

Studies have been undertaken of the forces generated in the muscles which move the human eyeball. The opportunity has been taken during operations for the treatment of squints to insert miniature tension transducers between the tendons and their points of insertion on the globe (Collins *et al.* 1975). There is always some tension in the muscles, and they never go slack. This may be due to tonic activity of the small outer fibres of the muscles.

Does the eyeball form a resonant system and, if so, what are its properties? Stone *et al.* (1965) thought they had demonstrated that the dog's eye resonated at 20 to 25Hz. The fallacy was that angular velocity and not displacement was recorded. For a flat frequency response, in terms of displacement, the velocity will rise with frequency (equation 6, p. 37) and will fall again once the displacement tails away at higher frequencies. This is not resonance!

Robinson (1964) investigated the human eyeball, applying a steady force to a contact lens which was held firmly in place by suction. When the tension was released the eye moved rapidly at first and then more slowly; the two phases of the motion had very different time constants. On the basis of a mathematical model, Robinson concluded that:

> The frequency at which the mechanical response of the eye and its attachments is 3 db below that at zero frequency is 1.1c/s.

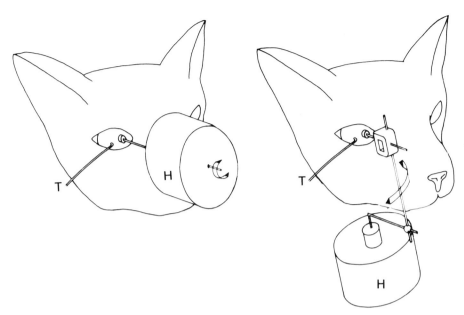

Fig. 7.15. *Left:* arrangement for generating torsional movements. *Right:* arrangement for horizontal motion. For vertical movements the same linkage was used as for horizontal movements but the apparatus was at right angles to that shown. The cylinder, H, is the housing for the motor and photocell; the motor itself was much smaller, having a diameter of 17mm and a length of 11mm. The drawings are not to scale: the plastic block in the linkage was smaller than portrayed.

Four of the parameters in Robinson's model were 'found by iterative approximations'. Is this the same as 'by guess and by God'?

As the discrepancy between the two studies is very great, the question was reinvestigated in association with J.A. Ashton, A. Boddy, I.M.L. Donaldson, M. Lakie and G.W. Wright. In these experiments, forces generated by a miniature basket-wound motor were applied to the eyes of anaesthetised paralysed cats via a contact lens held in place by suction (Fig. 7.15).

After premedication with xylazine, $0.1 mg.kg^{-1}$ intramuscularly, anaesthesia was induced with halothane 0.5 per cent increasing to 4 per cent in nitrous oxide/oxygen (3:1) delivered by face mask. The larynx was sprayed with lignocaine and a cuffed endotracheal tube inserted. The animal was mounted in a stereotaxic frame, a skull peg having previously been implanted; anaesthesia was maintained with sodium pentobarbitone 5 to $10 mg.kg^{-1}.h^{-1}$ as required. Neuromuscular block was with gallamine triethiodide $40 mg.kg^{-1}$ then $3 mg.kg^{-1}.h^{-1}$, and artificial ventilation was used. Body temperature was maintained at 38°C and tidal CO_2 was kept between 3.8 and 4 per cent. The ECG was monitored throughout. The motor (Maxon 2017938) was mounted on a micromanipulator. The static friction was $62 \mu N$ m. Observations under stroboscopic illumination showed that the contact lens remained locked solidly to the globe up to at least 90Hz. The motion of the globe was recorded photoelectrically. There was no macroscopic sign of damage or inflammation of the eye.

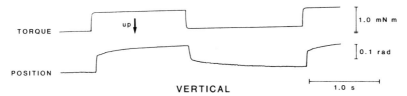

Fig. 7.16. Low-frequency alternating torque turning the eye alternately upwards and downwards. When the torque reverses there is at first a rapid movement of the eye followed by a motion having a much slower time course. There is a clear break in the curves when one motion gives way to another.

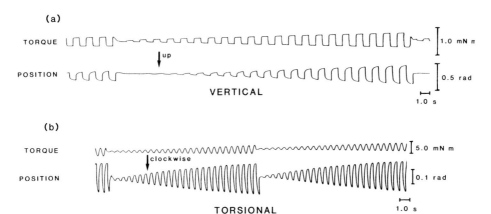

Fig. 7.17. *(a)* Vertical torques of low frequency are ramped in amplitude. The motion of the eye follows in an approximately linear manner. The rapid and slow components of the motion noted in Figure 7.15 are seen; the effect clearly does not depend on any particular position of the eye as it is seen over a wide range of amplitude of motion. *(b)* Torsional torques of sinusoidal form again ramped in amplitude; the relationship is again essentially linear.

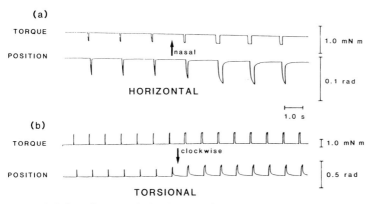

Fig. 7.18. Pulses of torque. *(a)* Horizontal. *(b)* Torsional. With sufficiently brief pulses the eye moves back to its initial position at once. With longer torques there is creep resulting in a somewhat greater amplitude and recovery after the force is withdrawn occurs in two phases.

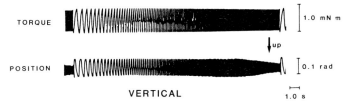

Fig. 7.19. Vertical movements. The forces applied to the eye are sinusoidal and increase in frequency logarithmically. No resonance is seen; the response is flat until it begins to fall off.

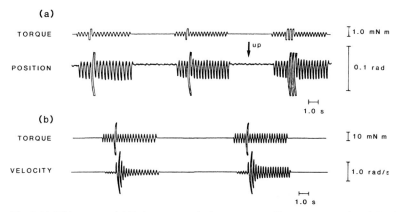

Fig. 7.20. Thixotropy. *(a)* Cat's eye—vertical movements. One or more cycles of the sinusoidal force are replaced by a larger rectangular perturbation. There is scarcely any increase in the amplitude of the motion after the disturbance. *(b)* Cat's ankle. First recording shows loosening of the system with a perturbation. Second recording—with a slightly higher force the effect is even more conspicuous. For this investigation a larger basket-wound motor (Phillips M010) was used. It was fitted with a tachometer. Upwards deflection—plantarflexion.

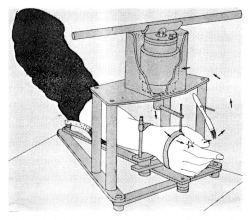

Fig. 7.21. 'Hanging hand tremorograph.' Tremor is recordable at rest, or the fingers may be extended to evaluate the effect of muscular activity. Inertia may be added, as shown, by positioning a bar of metal on the fitment at the top of the instrument. Perturbations may be introduced by striking the hand gently with a rotating brush. The motion was generated by a miniature geared motor (not shown); the rate of rotation of the brush was adjustable. (From Lakie *et al.* 1986*b*, by permission.)

Motion produced by low-frequency alternating torques
When the torque reversed there was a large and rapid movement of the eye followed by a much slower movement (Fig. 7.16). This phenomenon, where final equilibrium is slowly approached, may be the equivalent of the creep that is seen with some plastic substances. The results were similar for the three types of motion studied: torsional, horizontal and vertical. When the amplitude of the square waves was arranged to increase linearly, the creep phase was seen to be unrelated to any particular position of the eyeball (Fig. 7.17*a*).

Motion resulting from torque pulses
If a brief pulse was used the eye moved rapidly and then immediately returned to the starting position. With longer pulses the results were akin to those with rectangular torques. On the application of force the eye made a rapid movement followed by creep. When the force was over the eye flicked back but the original position was achieved quite slowly (Fig. 7.18).

Motion resulting from sinusoidal torques
In some experiments, sinusoidal forces of fixed frequency in the range 0.5 to 10Hz were arranged to increase linearly. The response was quasilinear (Fig. 7.17*b*), and again similar results were obtained for the three types of movement.

Similar results were found for the three types of motion when chirps were used (Fig. 7.19). There was no resonance, the motion being flat (within 1db) until it started to fall off at higher frequencies. The motion fell by 3db at 30Hz for torsional movements and at 17Hz and 22Hz for vertical and horizontal movements respectively. Alterations of the level of torques within wide limits did not materially change the resonances.

Thus the eyeball and its attachments do not form a resonant system, but equally the suggestion that the mechanical response falls off at a low frequency is quite false.

Search for thixotropic properties
Many different force levels and frequencies of sinusoidal torques were employed, and perturbations arranged as in the experiments on the wrist and other joints. No increase of motion greater than 10 per cent was ever seen (Fig. 7.20*a*). Almost no thixotropic effects could be observed in the eyeball experiments.

By contrast, clear thixotropic results were easily demonstrated at the ankle of the anaesthetised cat treated with a neuromuscular blocking agent (Fig. 7.20*b*).

Investigations of very small movements and the use of very small taps
The 'hanging hand' tremorograph
1. TREMOR IN THE RELAXED STATE
One set of observations was made using a 'hanging hand tremorograph' (Fig. 7.21). At rest a low level of physiological tremor was observed. At times it could be seen

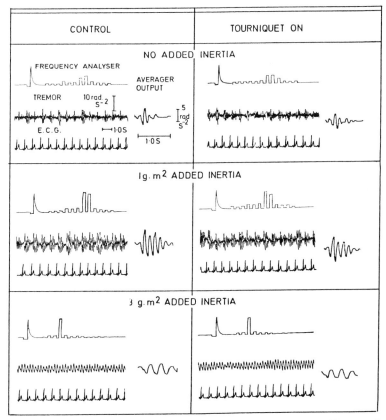

Fig. 7.22. Record of physiological tremor with hand in resting position together with a frequency analysis. The ECG, which is also displayed, has been used to trigger an averager. A significant proportion of the tremor is related to the pulse. In each example 16 sweeps have been averaged, each triggered by alternate R waves of the ECG. With added inertia the tremor becomes more sharply tuned and of lower frequency. The cardiac component too is slowed. When a tourniquet is applied to the upper arm the pulse-related component of the tremor persists. Frequency analyser pips from left to right correspond with the following frequencies (Hz): 2, 2.5, 3.1, 3.8, 4.6, 5.7, 7, 8.7, 10.7, 13.2, 16.2 and 20. The large exponentially decaying pulse indicated dynamic limit of analyser (from Lakie *et al.* 1983*b*).

that there was some modulation in time with the pulse. When the oscillations were averaged there was always a clear percentage locked to the cardiac cycle (Fig. 7.22).

When an inertia bar was placed on the top of the instrument, the rate of the oscillations dropped. Equation 12 (p. 75) indicates that added inertia increases Q. The 'raw' records of the tremor became more nearly sinusoidal and the frequency analysis sharper when the inertia was increased by $8g\ m^2$.

The relationship to the pulse remained when a tourniquet was applied to the upper arm to arrest the arterial circulation. At this time there was silence in the EMG channels. It was concluded that a significant part of the low level of motion observed was set up by the vibrations of the ballistocardiogram.

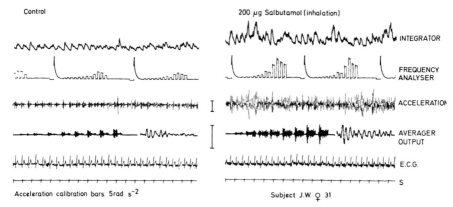

Fig. 7.23. Hand in resting position—influence of inhaling the bronchodilator salbutamol in a sensitive subject. Tremor increases. The top trace is from a leaky integrator connected to tremor signal (time constant 0.1s). The output of an averager triggered by the ECG for a 1s sweep builds up cycle by cycle and is displayed more slowly to the right of the trace. Contributions to the tremor of the ballistocardiogram increase through the action of the drug.

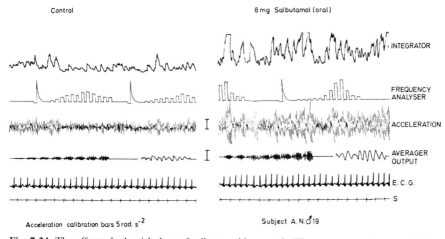

Fig. 7.24. The effect of a largish dose of salbutamol by mouth. The arrangements were similar to those of preceding figure.

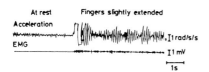

Fig. 7.25. When the fingers were partially extended the tremor amplitude was greatly increased. (From Lakie *et al.* 1986b, by permission.)

Observations were made using a beta-agonist drug, it being known that adrenaline increases physiological tremor. With large doses the tremor increased and the component locked to the ECG also rose (Figs 7.23, 7.24). It is thought that the drug will have increased the vigour of cardiac action, and much of the increase when the person is relaxed can be attributed to this cause.

2. POSTURAL TREMOR

When the fingers were partially extended from the position of rest, the tremor greatly increased and activity was seen in the EMG (Fig. 7.25).

With the maintenance of posture, or during a movement, tremor may be very obvious after the administration of adrenaline. Under these circumstances the contribution of the cardiac thrust is massively overshadowed by the vibrations caused by unfused muscular contractions. This effect was studied by Marsden *et al.* (1967), who found that the effect of intravenous adrenaline could be prevented on one side by the use of a tourniquet, or by the injection of a blocking drug, propanolol, into the brachial artery. With active muscular activity, the increase of tremor may be due to the changes of muscle properties known to occur and which have been reviewed by Rodger and Bowman (1983). There is a slowing of twitch time in fast muscle fibres and a speeding up in slow fibres. It is this shortening which is significant, for the slow fibres will be those concerned with holding a posture. The shortening of contraction time is modest, but with certain rates of excitation the ripple from unfused contractions ought to increase considerably, and the relationship between tremor and the change in muscle properties may be expected to be highly non-linear.

It is common knowledge that tremor may arise from strong emotions, and in these circumstances adrenaline is liberated from the adrenal glands. There are numerous references in the Bible to trembling from fear or anger, *e.g.* in Deuteronomy II. 25:

> This day will I begin to put the dread of thee and the fear of thee upon the nations that are under the whole heaven, who shall hear report of thee, and shall tremble, and be in anguish because of thee.

This association occurs in Shakespeare too, for instance in *Richard III* (I. iv. 58–62):

> With that, methought, a legion of foul fiends
> Environ'd me, and howled in mine ears
> Such hideous cries, that, with the very noise,
> I trembling wak'd, and, for a season after,
> Could not believe but that I was in hell.

2. THE USE OF SMALL TAPS

On applying a small tap to the suspended hand, a series of decrementing oscillations took place (Fig. 7.26). Their rate was a little faster when the fingers were extended instead of being in the resting position, and their frequency was lowered when inertia was added. From the change in the rate of oscillation of these

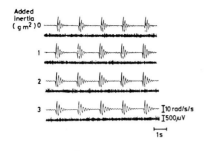

Fig. 7.26. Responses to taps with the fingers partially extended. Taps to the system induced transient oscillations at the same frequency as the tremor. There is tonic activity in the EMG but never any modulation even though the initial swings were much greater than the level of spontaneous tremor. There was thus no suggestion of stretch reflex activity. Added inertia slows the transients. (From Lakie et al. 1986b, by permission.)

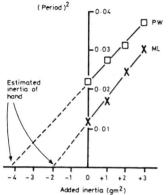

Fig. 7.27. Relationship between the frequency of the transients evoked by taps and the level of added inertia. As discussed elsewhere (see Fig. 6.5), the data can be used to determine the inertia of the hand. Results from two subjects are shown. PW had an unusually large hand. The extrapolation to zero indicates the inertia that would have to have been subtracted to reduce the period to zero. From the values the stiffness can be calculated in absolute units using equation 8, p. 49. (From Lakie et al. 1986b, by permission.)

transients it is possible to calculate the inertia of the hand and the stiffness of the muscles for this level of disturbance (Fig. 7.27). The procedure has been discussed elsewhere (see Fig. 6.5).

Measurements were made on patients being anaesthetised for abdominal operations and treated with a neuromuscular blocking agent. Thiopentone and halothane were the anaesthetic agents, suxamethonium the blocking agent. Of six patients tested, neuromuscular block was deemed complete in five. A low level of tremor persisted during the procedure and transients resulting from light taps were similar to those of normal relaxed people (Fig. 7.28).

The oscillations were thus certainly not of neural origin and represent a rather pronounced passive resonance. This confirms the findings that the resonance of the relaxed subject is basal (pp. 54–55). It has been noted earlier that the forearm is stiffer for small movements than for large ones. According to equation 12 (p. 75), Q ought to rise with increasing stiffness.

Taps produced by a miniature motor
An instrument which enables more accurately defined pulses of force to be employed is shown in Figure 7.29. In measurements of thixotropy at the finger, Lakie and Robson (1988b) used an apparatus which provided small taps. Each tap was followed by transient mechanical disturbances that took the form of a decrementing train of oscillations similar to those recorded with the hanging hand

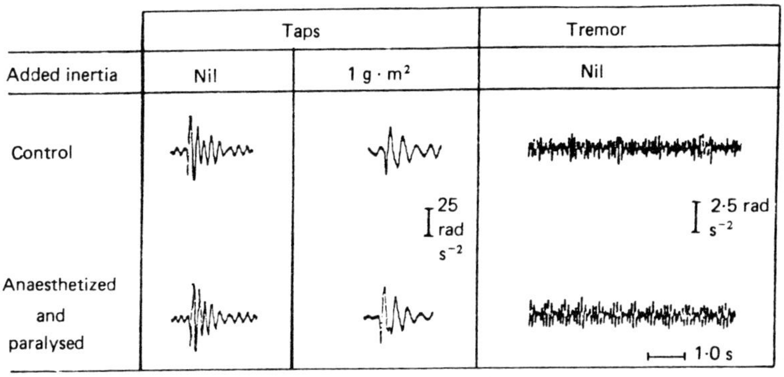

Added inertia	Taps		Tremor
	Nil	1 g·m²	Nil
Control	~^|^|^|^|^~	~|||~ $\mathrm{I}\begin{smallmatrix}25\\ \text{rad}\\ s^{-2}\end{smallmatrix}$	~~~~~~~ $\mathrm{I}\begin{smallmatrix}2{\cdot}5\text{ rad}\\ s^{-2}\end{smallmatrix}$
Anaesthetized and paralysed	~^|^|^|^~	~||~	~~~~~~~ ⊢—⊣ 1·0 s

Fig. 7.28. For small movements (1°) the wrist is quite stiff and significantly underdamped. The findings are essentially the same in anaesthetised paralysed patients. In this figure records were taken before and after the person was anaesthetised with thiopentone and halothane and treated with suxamethonium. The response to taps is unchanged both with and without added inertia. Low-level tremor persists (from Lakie *et al.* 1983*a*).

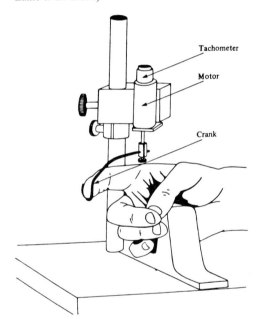

Fig. 7.29. A miniature basket-wound motor (Escap 23HD11) supplied small controllable taps to the finger to test stiffness. (From Lakie and Robson 1988*b*, by permission.)

instrument. This response was subjected to frequency analysis; the dominant frequency reflected the stiffness of the system, and from the time constant of the decay of the transients Q could be calculated. With this approach, following stiffening during a period of rest, there is minimal interference with the course of events.

As expected there was a large increase of stiffness in the first couple of seconds. Afterwards further stiffening occurred but the rate was slower; it appears that there are two time constants in the process. Thus, in addition to the short-term stiffening noted previously, there are longer-term effects that have not been

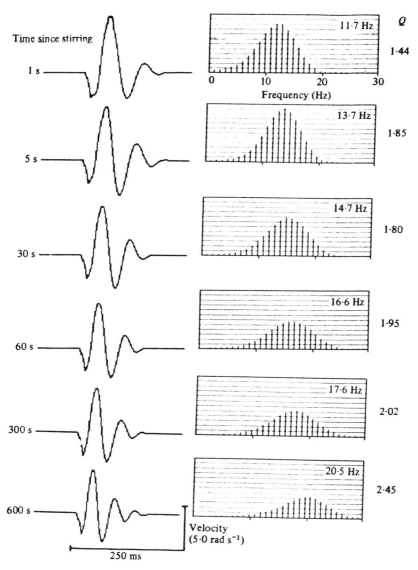

Fig. 7.30. *Left:* transients caused by taps at various rest times; there is progressive stiffening. The traces represent velocity, not displacement. *Right:* the peak of the power spectrum obtained by fast Fourier analysis increases from 11.7 to 20.5Hz after 10 minutes corresponding to an approximately threefold increase in stiffness. The sharpness of tuning of the transients (Q) was calculated; it increased as the stiffness rose. (From Lakie and Robson 1988b, by permission.)

extensively explored. Lakie and Robson (1988b) found progressive stiffening over a period of 10 minutes (Fig. 7.30).

Stiffening may be expected to rise during a period of undisturbed sleep. A cat, yawning and stretching, may be loosening up from such thixotropic stiffening.

Significance of thixotropy for physiotherapy
It is well known that when a plaster splint is removed the limb can feel very stiff. Limbs of patients with a variety of disturbances are often immobile for long periods and they are commonly cold. A number of the procedures used by physiotherapists may be expected to change postural thixotropy; thus passive or active movements, shaking, warming by wax baths or diathermy and probably also ultrasonics may all be expected to facilitate thixotropic loosening. It is possible that changes of a thixotropic type are important factors underlying the efficacy of physical medicine.

Significance of thixotropy for postural control
Thixotropy may be a way of eliminating small postural perturbations and reducing the need for the nervous system to react constantly. The role of the musculature is not only to impose movements but also to resist unpredictable externally applied forces; and, by making the load compensating mechanism the plastic properties of the muscle itself, inappropriate time lags and the instability that inevitably occur with feedback control are eliminated. Effectively the bones are embedded in a stiff jelly; the postural system is 'fixed in aspic'. A system that automatically stiffens when not in use, but which can be instantly loosened, may be of considerable value in postural stabilisation. If a posture is maintained, the EMG activity of the supporting muscles diminishes over a period of some seconds. It would appear that as the inactive fibres stiffen, a smaller proportion of the muscle mass needs to be activated to sustain the same load. For small movements the relaxed person is as rigid as a patient with a severe degree of Parkinsonism or with severe spastic hemiplegia following a stroke. The relative resistance of the postural mechanisms to fatigue has long intrigued physiologists and there is surprisingly little EMG activity in the standing position[1]. In postural sway the movements represent only a very slight elongation of the muscles concerned, so there is short-range stiffness. Furthermore, much of the activity is in the form of lengthening contractions, motor units being active but enlongating. In this situation a small amount of muscle will generate a comparatively high tension. Some animals have methods of locking their joints in the standing position (Walsh 1970c) and perhaps in man thixotropic gelation is important for standing.

The price to be paid for significant postural fixation by thixotropic stiffening is

---

[1]The role of muscle tone in maintaining the arch of the foot was the subject of a study by Basmajian and Steko (1963). They found that loads of up to 200lb could be supported by the ligaments and bones without muscular activity as recorded by the EMG. Only heavy loading (e.g. 400lb) caused significant activity.

that motor impulses will be more or less effective in generating force according to the past history of movement. For the eyeball alone gravity is of virtually no consequence. As noted above, the eye muscles do not behave like this. As the eyeball is virtually balanced with respect to gravity it is difficult to see what biological merit there would be in the existence of significant thixotropy, and the control of eye movements by the nervous system would be complicated. The programming of eye movements by the nervous system may well be more straightforward for this system where thixotropic stiffening is evidently trivial. Somehow the nervous system must take into account this apparently anarchic behaviour when planning and executing movements of the limbs.

8
HYPOTONIA

The methods described in earlier chapters enable estimates to be made of the deviations of stiffness from normal values and in certain circumstances, tone is reduced below normal. Some of these are considered in this chapter.

The disconnected terminal phalanx
You can experience a state of hypotonia in your own hand by positioning the fingers appropriately (Fig. 8.1). An anatomical peculiarity enables the tendons from both the flexor and extensor groups to become so slack that the muscles no longer control the terminal phalanx. The anatomy underlying this phenomenon has intrigued anatomists (*e.g.* Landsmeer 1949) and has been of interest to physiologists trying to assess the relative contributions of joint *vs* muscle receptors to proprioception (Gandevia and McCloskey 1976). Position sense is seriously impaired when the muscles are functionally disconnected. The biomechanics of this situation have not been extensively explored. Barnett and Cobbold (1969) connected the finger to a heavy pendulum and measured the decay in the amplitude of the motion after the system was started swinging; this was a reflection of damping, and it was reduced to less than half when the phalanx was disconnected. Their rather crude method could give no information about stiffness.

Reference has already been made to the observations of Lakie and Robson (1988*a*) that no thixotropy was observable in this position (p. 89).

Floppy infants
The floppy infant has been likened to a rag doll; bizarre postures are adopted, and there is diminished resistance to passive movements and an increased range of passive movement. Floppiness is common in infancy and may be associated with a very wide variety of apparently unrelated conditions. In a monograph on this subject, Dubowitz (1980) related that:

> It may be the presenting feature of a neuromuscular disorder; it may occur in mentally retarded children or in the early phase of cerebral palsy; it may be a manifestation of a connective tissue disorder; it may be associated with various metabolic disorders in infancy; it may be an incidental and non-specific sign in any acutely ill child; it may be completely physiological in the premature infant; and it may occur as a completely isolated symptom in an otherwise normal child.

Shufflers, creepers and rollers
Some infants with hypotonia dislike lying prone and when held upright usually flex their hips and extend their knees so that the lower limbs are held out in front of them (Robson 1970). These children move around in a sitting position in a manner known as 'shuffling', 'hitching' or 'scooting'. It is a sliding action with the trunk

Fig. 8.1. *(a)* In this position with the other fingers extended the terminal phalanx of the middle finger cannot be used for the muscles controlling the joint are slack and inoperative—there is 'uncoupling'. *(b)* With the other fingers flexed control is re-established. (From Barnett and Cobbold 1969, by permission.)

Fig. 8.2. Buddha in 'lotus position'—soles of feet upwards. Not a position adopted as a result of religious asceticism but no doubt natural; hypermobility is relatively common in orientals.

erect and the hips flexed, the weight being taken on the buttocks or on one buttock and a flexed leg. The feet or hands or both may be used for propulsion. They are late in standing and walking and there are often many relatives who have shuffled. Robson (1984) gave this account of prewalking locomotor movements:

> The majority (82%) of normal infants crawl on hands and knees as the predominant means of moving from place to place before they get themselves to standing. Others shuffle in a sitting position (9%), creep on the abdomen (1%) or roll (1%), and tend to walk much later than the crawlers. The earliest walkers have no observable pre-walking locomotion—they just stand up and walk (7%). In many instances, the age at which one locomotor milestone is attained correlates well with the age at which subsequent milestones appear, thus permitting prediction of the age of standing and walking. Such predictions are useful in offering parents and therapists a time scale over which future skills can be expected to develop in both normal and handicapped children.

It has been suggested jocularly that some names (such as Shufflebottom, Sidebottom, and perhaps Botham, Bottomley, and Higginbottom) referred to infants that progressed in this way or perhaps to families with this trait. This seems untrue: according to appropriate dictionaries these were originally place names referring, to, for instance, a valley bottom.

Hypermobility

Limb flexibility varies considerably from person to person. Some people are 'double-jointed' or hypermobile. Only a minority of people can sit in the lotus position with the soles of the feet facing upwards, though hypermobility is more common among Africans and Orientals than Europeans (Fig. 8.2). It is hard to say to what extent this is genetic. In an Indian village there may be few chairs, so most people squat on the ground and the range of posture to which they are accustomed from early childhood is different. Thai dancers make use of the hypermobility of the joints of their hands.

There are skeletal differences according to the postures customarily adopted. Rubin (1974) wrote:

> An interesting development which illustrates one feature of early Anglo-Saxon social life is the common appearance of the 'squatting facet' on the lower end of the tibia. This facet develops after prolonged adoption of the squatting position and is found less frequently in later Saxon times, suggesting that better living standards were being enjoyed and the use of chairs and benches was becoming more common. At Little Eriswell, Suffolk, for instance, no less than seven out of ten individuals show evidence of squatting facets at the ankle joint. This high proportion must indicate poor economic and domestic standards. At North Elmham, Norfolk, of a group of 206 persons dating from 950 to 1050 (A.D.), as many as 66 (80.0 per cent) of the women and 34 (40.5 per cent) of the men display squatting facets, indicating not only that the group were living in relative poverty but that the women endured a ruder, less comfortable existence than the men, and that whatever comforts may have been available, the men had first claim. Possibly this high proportion of female facets was due to the habit of squatting for long periods weaving cloth and performing other crafts and duties.
>
> An additional feature resulting from habitual squatting is the extension present on the articular surface of the medial condyle of the femur on its posterior surface. Long periods of squatting with the resulting extreme degree of knee flexion would be enough to produce this articular elongation.

If the range of possible motion is sufficiently great, the person is said to have the benign hypermobility syndrome. A series of simple tests can be undertaken without special equipment (see Fig. 8.5 and review by Beighton *et al.* 1989). The range of movement of a given joint in different individuals is distributed in a Gaussian manner.

The benign hypermobility syndrome is usually said to be due to 'joint laxity'. Hypermobile people are at one extreme of a normal distribution, although hypermobility is very occasionally due to pathological conditions. There are hereditary disorders of connective tissue (*e.g.* Marfan syndrome, Ehlers-Danlos syndrome). These conditions are associated with other abnormalities and it is this association which leads to an appropriate diagnosis.

Hypermobility is about twice as common in females as in males. The range of normal joint movement decreases rapidly throughout childhood and more slowly in adult life. Hypermobile people are prone to a variety of musculoskeletal complaints: muscle pains, dislocation of the patellae, joint effusions, bicipital tendinitis, tennis elbow, Baker's cyst, chronic back pain, disc prolapse and spondylolisthesis. Some hypermobile people dislocate their shoulders as a party trick; this is foolish.

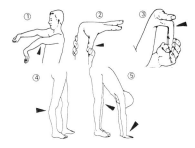

Fig. 8.3. Tests for hypermobility. 1. Can either of the wrists be hyperextended? 2. Can either of the thumbs be pressed onto the forearm? 3. Can the fingers of either side be hyperextended? 4. Do the knees on either side hyperextend? 5. Can the person touch the floor with the palms of the hands? If all of the tests are positive a score of nine is obtained, if none the score is 0. A score of 5 of more is regarded as indicating hypermobility. (From Rowe and Shapiro 1989, by permission, after Wynne-Davies.)

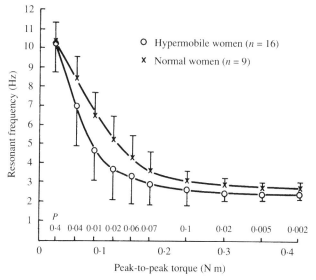

Fig. 8.4. Relationship between applied torque and resonant frequency at the wrist in normal and hypermobile women. At the lowest torque the results are identical but the other values for the hypermobiles are lower, indicating decreased stiffness. The p values on the abscissa refer to the points above and were obtained by the use of Student's t test. For quite modest movements the hypermobiles are less stiff (from Walsh *et al.* 1991).

Nicolo Paganini (1782–1840), the violin virtuoso, possessed an extraordinary facility with his left hand which became the subject of intense inquiry as his career progressed (Smith 1982). He could move his joints laterally and bend the thumb back until it touched the little finger. He moved his hand as flexibly as though it were without muscle or bones. He could span three octaves with ease, taking in one bow four Cs, or four Ds, or four E flats in the octave across the four strings by using the first, second, third and fourth fingers. He could bring the thumb of his left hand over the middle of the fingerboard in such a way that he could play, at pleasure, in the first three positions without the 'extension'. He probably suffered from Ehlers-Danlos syndrome.

It was decided to see whether the resistance to passive movements was reduced for motion of moderate amplitude and, if so, to consider the implications. A score of 5 or more in the tests for mobility (Fig. 8.3) was taken as evidence of hypermobility. The subjects (all women, as female subjects could be found more easily) also answered a questionnaire, devised by Dr M. Lambert, and were examined medically to exclude those with a pathological condition. Measurements of the resonance at the wrist at 10 torque levels were compared with those of an age-matched group (Fig. 8.4). In the hypermobiles as in the normals the resonant frequency rose at the lower force levels. Except at the lowest force level, the resonant frequencies of the hypermobiles were all below those of the normals, indicating less stiffness. There was a clear and statistically significant difference even when the motion was quite small and nowhere near the anatomical limits. There was no difference in thixotropy or grip strength, and therefore no evidence that the contractile mechanisms were abnormal.

Joint motion is restrained partly by the muscles and partly by the ligaments of the joints. In the dissecting room, when the tendons of the wrist are divided, there is almost no resistance to motion until the extremes of movement have been reached, when the ligaments tense. In anaesthetised sheep, the author has also observed that when the tendons at the corresponding joint are divided there is virtually no resistance to movement until extreme flexion or extension is approached; the joint is flail-like.

It was concluded that the muscles in hypermobility are less stiff than normal and that this contributes to the laxity. The braking effect of the ligaments of the joints is also less pronounced. The resistance to stretching a complete muscle depends partly on the properties of the muscle fibres and partly on the connective tissue, endomysium and perimysium. Connective tissue itself is not compliant but the arrangement of the fibres can provide elasticity. There is a meshwork of connective tissue whose fibres become taut when the muscle is stretched, as with a knitted stocking. It is likely that hypermobile subjects have less (or different) connective tissue in their muscles. This suggestion is compatible with the known reduction of hypermobility with age, mobility being greatest in infancy (mutton is tougher than lamb!) In *Hamlet* (III. iii. 71–2), Claudius says:

> Heart with strings of steel,
> Be soft as sinews of the new-born babe.

It is likely that in the relaxed limb the resistance to motion of limited extent is determined by cross bridges in the muscle fibres, *i.e.* thixotropy: if the motion is moderate by the connective tissue of the muscle, and if large by the ligaments of the joints.

The paradox of paraplegia

Following a spinal injury, a person may develop clonus, suffer from spasms and, commonly, have exaggerated reflexes. S/he is thus said to suffer from spasticity. It was in the expectation that muscle stiffness would be increased that arrangements

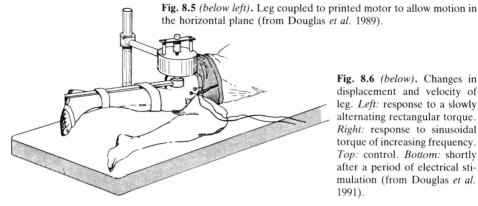

Fig. 8.5 *(below left).* Leg coupled to printed motor to allow motion in the horizontal plane (from Douglas *et al.* 1989).

Fig. 8.6 *(below).* Changes in displacement and velocity of leg. *Left:* response to a slowly alternating rectangular torque. *Right:* response to sinusoidal torque of increasing frequency. *Top:* control. *Bottom:* shortly after a period of electrical stimulation (from Douglas *et al.* 1991).

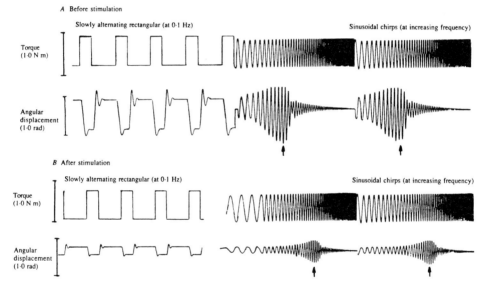

were made to measure the resonant frequency of the knee. The apparatus is shown in Figure 8.5 and a specimen recording in Figure 8.6. It was found that in normal subjects the resonant frequency rose as the torque was reduced. Most normal subjects tested were able to relax fully, as shown by the consistency of the recordings and the absence of electrical activity in the thigh muscles. It was a surprise to find that the paraplegics' muscles were also generally silent both at rest and with movements of low or moderate velocity. With the larger torques at resonance and consequent wide swings there was some activity, evidently due to phasic stretch reflexes. Naturally when the patient had a spasm there was considerable electrical activity in the muscles, and spasms were sometimes provoked by the larger wider swings. For the moderate and larger torques used there was no difference in the biomechanical parameters between the normal and the paraplegic subject. The latter, however, did not show the rise of resonant

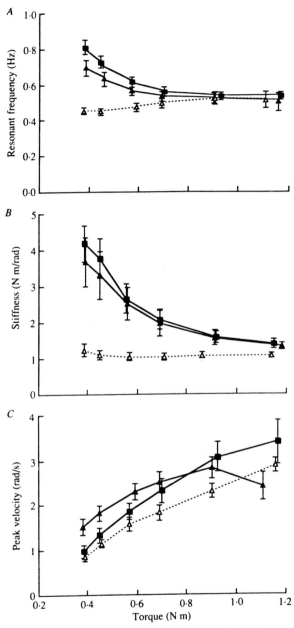

Fig. 8.7. Measurements at the knee for 14 controls *(black squares)*, 14 paraplegic subjects not treated by electrical stimulation *(hollow triangles)* and five paraplegics treated by electrical stimulation *(black triangles)*. Mean values ± SEM. *A:* Tests with chirps to obtain resonant frequency. Patients who have had electrical stimulation have a more nearly normal graph with the rise of resonant frequency at the low torques. This is absent in the untreated paraplegics over the range tested. *B:* stiffness calculated from data with low frequency square waves. *C:* velocity at resonance, damping is broadly similar in the three groups but the untreated paraplegics values are somewhat lower (from Douglas *et al.* 1991).

TABLE 8.I
Day-to-day variability of resonant frequency, stiffness and peak velocity in controls and paraplegic subjects
Day-to-day analysis of resonant frequency, stiffness and peak velocity at the knee of normal subjects

Day	Subject					
	1	2	3	4	5	6
	Resonant frequency (H_2)					
1	0.54 ± 0.01	0.67 ± 0.01	0.43 ± 0.01	0.42 ± 0.01	0.63 ± 0	0.69 ± 0.01
2	0.57 ± 0.01	0.63 ± 0.01	0.35 ± 0.01	0.43 ± 0.01	0.53 ± 0.01	0.63 ± 0.01
3	0.51 ± 0.01	0.44 ± 0.01	0.39 ± 0.01			
	Peak velocity (rad/s)					
1	1.51 ± 0.02	1.39 ± 0.01	1.77 ± 0.01	2.09 ± 0.1	1.80 ± 0.01	1.90 ± 0.01
2	1.55 ± 0.02	1.54 ± 0.01	1.50 ± 0.02	2.22 ± 0.12	1.77 ± 0.04	1.68 ± 0.01
3	1.69 ± 0.07	1.39 ± 0.02	1.87 ± 0.01			
	Stiffness (N m/rad)					
1	2.18 ± 0.16	3.6 ± 0.11	1.81 ± 0.04	1.02 ± 0.04	1.98 ± 0.11	2.74 ± 0.1
2	2.13 ± 0.19	3.1 ± 0.08	2.49 ± 0.04	1.42 ± 0.13	1.62 ± 0.14	1.87 ± 0.03
3	1.83 ± 0.05	2.81 ± 0.11	1.50 ± 0.01			

Values are mean ± SEM of five consecutive measurements

TABLE 8.II
Day-to-day analysis of resonant frequency, stiffness and peak velocity at the knee of complete-lesion paraplegics

Day	Subject				
	1	2	3	4	5
	Resonant frequency (Hz)				
1	0.42 ± 0.01	0.51 ± 0.02	0.36 ± 0.01	0.47 ± 0.01	0.40 ± 0.01
2	0.54 ± 0.01	0.56 ± 0.01	0.33 ± 0.01	0.54 ± 0.01	0.38 ± 0.01
3	0.4 ± 0.01	0.48 ± 0.01	0.31 ± 0.01		
4	0.4 ± 0.01				
	Peak velocity (rad/s)				
1	2.89 ± 0.05	3.79 ± 0.1	4.20 ± 0.02	3.94 ± 0.08	3.73 ± 0.03
2	2.04 ± 0.02	3.63 ± 0.06	3.94 ± 0.04	1.61 ± 0.02	3.15 ± 0.03
3	2.93 ± 0.01	3.93 ± 0.06	4.10 ± 0.02		
4	2.52 ± 0.08				
	Stiffness (N m/rad)				
1	3.15 ± 0.21	1.37 ± 0.06	0.59 ± 0.03	1.43 ± 0.06	0.87 ± 0.01
2	2.02 ± 0.09	0.88 ± 0.01	0.87 ± 0.04	1.69 ± 0.02	0.95 ± 0.05
3	2.68 ± 0.06	1.30 ± 0.03	0.73 ± 0.01		
4	2.34 ± 0.08				

Values are mean ± SEM of five consecutive measurements

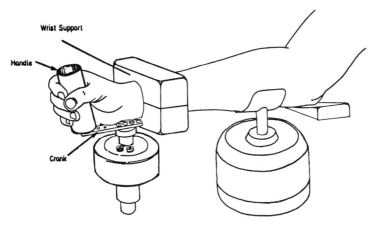

Fig. 8.8. Printed motor to apply torques to the wrist and a moving coil instrument to apply controlled vibration to the forearm. The vibrator used was model V 406 made by Ling Dynamics Systems Ltd, Royston, Herts, SG8 5BQ.

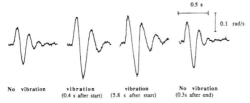

Fig. 8.9. Small pulses of force applied to the wrist provoked oscillatory transients. Substantial increase with vibration. Pulse duration 0.08s, strength 0.2Nm in flexor direction. Vibration of 3.3mm at 30Hz GWW, male, aged 53 years.

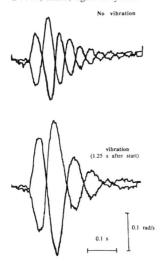

Fig. 8.10. Small pulses of force applied to wrist—oscillatory transients before and during vibration. The results of using flexor direction pulses and extensor direction pulses are mirror images—the records have been superimposed. Flexion upwards. There is a large increase in the motion during the vibration. Pulse duration 0.08s, strength of 0.2Nm, Vibration of 3.3mm at 30Hz, GWW, male, aged 53 years.

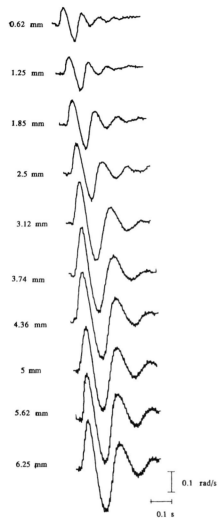

Fig. 8.11. Small pulses of force applied to wrist—oscillatory transients—effect of increasing amplitudes of vibration. Pulse duration 0.08s, strength 0.05Nm, direction flexor. Records 0.7s after start of vibration, frequency 30Hz, amplitudes on left of records. NP, male, aged 55 years.

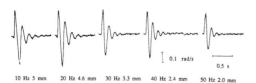

10 Hz 5 mm   20 Hz 4.6 mm   30 Hz 3.3 mm   40 Hz 2.4 mm   50 Hz 2.0 mm

Fig. 8.12. Small pulses of force applied to wrist—oscillatory transients—effect of varying frequency. As the velocity of an oscillation is proportional to frequency (equation 6, p. 39) it increased although the amplitude went down as the rate was increased. The motion was greatest at 10Hz and 20Hz. Pulse duration 0.08s, strength 0.05Nm, direction flexor. Records 0.1s after start of vibration. NP, male, aged 55 years.

frequency with the lower forces (Fig. 8.7); in patients in whom the muscles had been subjected to a course of electrical stimulation, the curve was nearer normal.

The opportunity was taken to measure day-to-day variability in both normal subjects and paraplegics. The 'stiffness' (N m/rad) was measured by using low-frequency alternating forces (see Chapter 9). For both resonant frequency and stiffness measurements a torque of 0.7N m was used. The differences were mostly small, as was the variation (Tables 8.I, 8.II). People with spinal cord injuries moved more for a given force than control subjects, and no evidence of increased tone was found at the knee in paraplegia.

Vibration-induced hypotonia
Vibration is used in physiotherapy departments to treat stiffness and relieve muscle pains. Many of the vibrators sold for this purpose run at twice the mains frequency, but it would be surprising if 100Hz (120Hz in the USA) happened to be the optimum frequency for use.

It was accordingly decided to investigate the effects of vibration of the forearm on muscle tone measured at the wrist. The observations were undertaken in association with Dr M. Lakie, and the apparatus is shown in Figure 8.8. Stiffness was tested by the use of pulses of force. With small forces provided by the motor it was seen that the motion increased by vibration. The changes were often substantial and rapid, and the decrease of tone occurred almost as soon as the vibration was applied (Fig. 8.9). They occurred equally for pulses of force in the flexor and extensor direction (Fig. 8.10). Over the range studied the looseness increased with the level of vibration (Fig. 8.11). In terms of the velocity or acceleration produced by the vibrator it was clear that low frequencies were the most effective (Fig. 8.12). The stiffness was regained very rapidly after the vibration. The resonant frequency fell, as determined by the use of swept sine waves; the damping too was reduced. There was no tendency for the loosening to be self-perpetuating when the limb was kept in motion by sinusoidal torques from the motor. Thus this form of loosening is distinct from the thixotropic effects described earlier. There were no associated EMG discharges; the changes are in the passive properties of the tissues.

Vibration can contrariwise cause muscle stiffening and such effects may be widespread in the animal kingdom. Alkon (1987) studied the response of a mollusc (Hermissenda) to rotation and shaking as would occur when the creature is in a turbulent ocean. The animal retracts its foot and can thus adhere to the surface on which it moves. The response can readily be conditioned so that the contraction occurs on illumination. This alteration in behaviour was shown to reside in the membrane properties of certain neurons.

If the tendon of a muscle is vibrated at 100Hz, for instance, there is a contraction, the tonic vibration reflex, which takes a few seconds to build up. This response is due to excitation of primary muscle spindles and has been studied in the decerebrate cat by Matthews (1966). On increasing the vibration at any particular

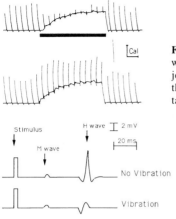

Fig. 8.13. *Top:* effect of vibration on patellar reflex. The jerks were elicited at 5s intervals. There is a tonic contraction and the jerks are suppressed. *Bottom:* comparable voluntary contraction, the jerks are not suppressed. Calibration vertical 0.6kg, horizontal 10s. (Adapted from Gail *et al.* 1966, by permission.)

Fig. 8.14 *(left).* Inhibition of H-reflex by vibration at 80Hz, of 2mm, a vigorous motion. Loosely based on the work of Desmedt and Godaux (1978).

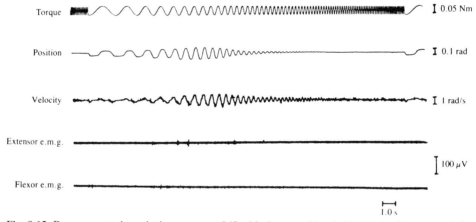

Fig. 8.15. Resonance at the wrist in a woman of 63 with rheumatoid arthritis. As the frequency of the applied torque increases the motion goes through a maximum. The EMG of both the extensors and the flexors is virtually silent even at high gain—relaxation is satisfactory (from Walsh *et al.* 1989).

frequency in the range 100 to 200Hz the reflex tension increased to a plateau for amplitudes of 100 to 200μm. Further increase of amplitude decreased the tension. As there was no concomitant decrease in the EMG it was evident that in addition to its reflex effects vibration can change the contractile properties of the muscle directly. But the reflex effects of vibration are complex: in addition to spinal excitation presynaptic inhibition may occur, and the overall response depends on the balance between the processes. Tendon jerks can be reduced by vibration (Fig. 8.13). The jaw jerk is exceptional: it is not decreased. Presynaptic inhibition has been shown to decrease and may be a significant cause of spasticity. Vibration can also inhibit the H reflex (Fig. 8.14). One review of this work is that of Whitlock (1990). Other physiological effects of vibration are discussed in Chapter 11 (pp. 174–181).

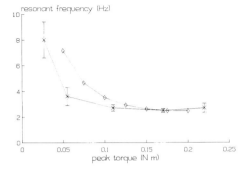

Fig. 8.16. Relationship between resonant frequency and peak to peak torque in seven women patients with rheumatoid arthritis *(crosses—means ± SD)* and eight normal women *(diamonds)*. In the arthritis the curve is shiften to the left (from Walsh *et al.* 1989).

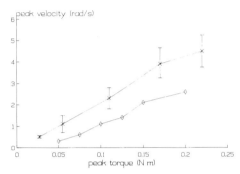

Fig. 8.17. Relationship between peak velocity and peak-to-peak torque in arthritics and controls. Symbols as in Figure 8.16. Higher velocities are reached in the arthritics (from Walsh *et al.* 1989).

The enigma of stiffness in rheumatoid arthritis

A feeling of stiffness is a cardinal symptom of rheumatoid arthritis, and it has been argued that an objective measurement would be of considerable value in discriminating between inflammatory and degenerative joint disease and in assessing the value of different treatments. Previous work in this field had given discordant results, and it was decided to measure the stiffness and damping at the wrist using the resonant frequency method described in Chapter 4. The records obtained were similar in their main features to those of normal subjects (Fig. 8.15). When resonant frequency was plotted against applied torque, however, quantitative differences emerged (Fig. 8.16). For moderate and small forces the arthritic patients had lower resonant frequencies than the controls. This indicates less rather than more stiffness. There is some wasting of the hand and shortening of the fingers with osteoporosis, so the inertia will be lower; however there is also muscle wasting. Whatever arguments may be based on these considerations, there is no enhanced stiffness in relation to the existing inertia because the resonant frequency was not elevated.

For the measurement of damping, inertia is irrelevant since inertial forces at resonance are balanced by those due to stiffness. The velocity at resonance in relation to the applied torque is shown in Figure 8.17. Damping was, in absolute terms, less in the patients. As for thixotropy, there was no statistically significant difference between the patients and the controls.

The study was started in the expectation that the subjective feelings of stiffness experienced by patients with rheumatoid arthritis would be reflected in biomechanical changes. This expectation has not been fulfilled. A given torque generally causes more movement in rheumatoid arthritis than in normal subjects, although the overall range of movements is restricted.

Cataplexy
This condition is defined as an abrupt decrease in muscle tone, either generalised or limited to particular muscle groups. Muscle activity is abolished and tendon jerks are in abeyance. The person is conscious throughout, and attacks may persist for a lifetime. There is no change in the EEG, although with prolonged attacks sleep may supervene. A similar condition is known to occur in certain breeds of dogs (Kushida *et al.* 1985). Nothing seems to be known about the relevant disturbances of physiological function but attacks may be precipitated by emotion—surprise, exultation, anger or laughter. Everyone is aware that fear may make them 'weak at the knees', and perhaps cataplexy is an exaggeration of this normal response.

Drop attacks
This subject was reviewed admirably by Lipsitz (1983). Sheldon's (1960) account of the typical attack was of a sudden, unexpected fall to the ground often after turning the head or neck. There is no loss of consciousness. The loss of strength may last several hours. Injuries are common as a result of the fall. One cause may be ischaemia of the brainstem caused by compression of the vertebral artery consequent on neck movements. Although principally a problem of old age, one study showed that it is not uncommon in middle-aged women. Rather strangely, muscle tone is said to be restored by pressure on the soles of the feet.

The cerebellum
The cerebellum plays a major role in regulating muscle tone and the classic work is by Gordon Holmes (1917). He was then neurologist to the British Expeditionary Force in France. As a staff officer he had a kymograph in the back of his open car and made his observations in difficult conditions near the front:

> The opportunity of making uncomplicated clinical observations is rare in civil life, since acute lesions of the cerebellum, comparable with those produced by physiologists, are uncommon; tumours and abscesses which develop in it are very liable to compress or influence the functions of other parts too, and the degenerative and atrophic diseases which involve it practically never affect it alone. In warfare on the other hand wounds limited to the cerebellum and injuries of it of different extent and localisation can be frequently observed.

He observed the swinging of the legs after the knee jerk had been elicited. With a unilateral lesion there were more oscillations on the ipsilateral side, and he noted that the muscles on that side felt flabby. When the arm was shaken, the hand swung in a flail-like manner. There were other changes too:

> If the fingers are seized and passively extended at the same time as the wrist they can easily, and without the observer experiencing any resistance, be bent back till further movement becomes

> impossible owing to the conformation of the joints and the tension on their ligaments; and even in this position the patient does not experience the dull pain or discomfort in the overstretched flexor muscles which he suffers when the unaffected wrist and fingers are forcibly extended to the same degree.

Holmes observed that changes of this type could occur very rapidly after an acute destructive lesion of the cerebellum, with some further increase in the signs over a period of 10 days or so. It seems likely that protective reflexes are diminished or in abeyance, and over a period of some days there may be physiological changes in the musculature. The state of the muscles is heavily dependent on the discharges which reach them, and, if the innervation changes, muscle properties alter. In a review, Salmons and Henriksson (1981) stated that:

> In spite of their high degree of specialization mammalian skeletal muscles have a remarkable capacity for accommodating changes in demand. In the process they acquire physiological and biochemical characteristics which appear suited to the new functional requirements.

Perhaps the hypotonia of cerebellar lesions in man is related to the changed innervation.

Conclusion

Hypotonia one way or another is common. It may be physiological or associated with a wide variety of pathological conditions. In coronary thrombosis, for instance, the person often complains of weakness in the arms and, when manipulated, the feeling is of floppiness. Very few cases of the hypotonic conditions have been investigated with appropriate modern methods.

9
THE USE OF ABRUPTLY ALTERNATING TORQUES TO INVESTIGATE THE POSTURAL SYSTEM

So far in this book, most of the laboratory observations have been concerned with the use of sinusoidal torques, feedback-induced motion and short pulses of force. In investigating control systems, engineers appreciate that useful information can often be obtained by an alternative system—the use of step functions. With appropriate instrumentation it is possible to change from sinusoids of adjustable frequency, to chirps, to positive velocity feedback, to pulses, to random noise, or to slow rectangular waves by a multiway switch. It is thus possible to 'interrogate' the postural system in a wide variety of ways. This multiplicity is valuable in documenting a system with complex properties.

The purpose of this chapter is principally to describe the use of a procedure where low-frequency abruptly changing torques are used and the changes of position and the velocity of the resultant motion are recorded together, where appropriate with the EMG.

As has been seen the resonant frequency method gives information related to stiffness but special procedures are needed to obtain absolute values. With low-frequency alternating torques it is possible to obtain absolute values for stiffness but the relative contribution of thixotropy can be inferred only by the type of study described in earlier chapters.

Observations on relaxed and anaesthetised subjects
Torques were applied to the wrists of normal relaxed subjects alternating between flexion and extension at about 1Hz. As the torque reversed, the hand moved at first but shortly afterwards came to rest in a new equilibrium position. Before the new position was achieved there was often some overshoot. Such responses show that the system is very slightly 'under-damped'.

In a system which has mass and elasticity, the response to a disturbance is a function of the damping. If damping is light there will be 'ringing'; if there was no damping the oscillations would continue indefinitely. If damping is heavy the system will sluggishly move to equilibrium. With infinite damping the system would never move. With a forcing step function, such as is considered in this chapter, the new position will be reached more rapidly with very light damping if the damping is increased, and with very heavy damping if it is decreased. There is an optimum for speed of response where there is a little overshoot, a condition known as critical

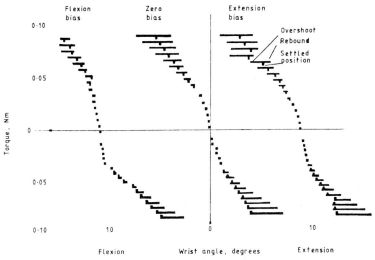

Fig. 9.1. Composite compliance curves for the wrist. Torque and displacement are recorded by FM tape recorder and replayed together at slow speed onto an X-Y plotter. The curves show a stiff zone for small forces and considerable overshoot for the larger forces. When flexion or extension bias torques are applied the stiff zone is displaced to the new position of rest (from Lakie *et al.* 1979*a*).

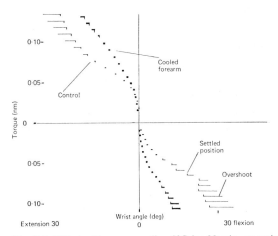

Fig. 9.2. Effect of forearm cooling 2°C for 20 min. on wrist compliance using the tank method. Female aged 22 (from Lakie *et al.* 1979*b*).

damping. Very often the postural system has been found to have properties approximating to this type.

With torques of varying strength, the wrist was stiffer for small than for large forces. When a steady bias was applied to the motor in a flexor or extensor direction, the stiff region was displaced to a new position (Fig. 9.1). These results are readily explicable on the basis of postural thixotropy (Chapter 7). When the forearm was cooled by the tank method (Fig. 7.4) there was a clear alteration of the compliance curves (Fig. 9.2).

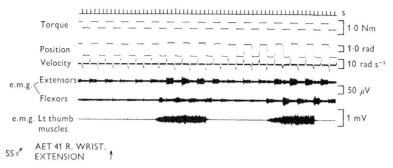

Fig. 9.3. Shortening reactions. It is often supposed that in man owing to the operation of stretch reflexes muscles contract when pulled upon. In the relaxed man electrical activity has been repeatedly seen not in the muscle being stretched but in the muscle that is being unloaded. About 100ms after the tension is removed from a muscle it may become active. The response is increased when the person grips powerfully with the opposite hand *(lowest trace)*. The amount of activity varies considerably with finger position. One subject capable of deliberate relaxation could abolish the activity without change of posture—there was almost no change in the biomechanics. By taking in the slack in an alert subject the muscles are prepared for action (from Walsh 1976a).

Observations were made on three patients being anaesthetised and given full doses of relaxing drugs for laminectomy operations. There appeared to be no difference between the biomechanics of the wrist in the conscious and the anaesthetised states.

It has been a commonly accepted dogma that the muscles respond to stretch through reflex activity. The evidence was reviewed by Ralston and Libet (1953); the data showed that the minimum requirements, in terms of rotation of the appropriate limb portion, were about 45° in 0.2s for the anterior tibial and quadriceps femoris and somewhat less for the soleus. These are high rates of stretch, beyond the limits normally encountered in most city dwellers. They are more comparable to the levels reached in fighting. Using the apparatus already described (Fig. 4.5), stretch reflexes were never seen in the many thousands of normal relaxed subjects who were observed.

For the resting subject the stretch reflex may be principally a protective reflex; it comes into play only at high levels of excitation. In movement the system may spring to life, and the fusimotor neurons which bias the muscle spindles then become active (Burke 1980).

Observations on normal subjects with varying degrees of voluntary stiffening
In a person who is fully relaxed mentally and physically there is commonly no EMG activity in the flexor and extensor recording channels. In one who is somewhat less relaxed, shortening reactions are seen: there is activity in the extensor muscles when the hand is pushed into extension, and activity in the flexor channels when the hand is pushed into flexion. Shortening reactions are also seen with sinusoidal torques (pp. 53–54). They can be increased when the person uses muscle in another part of the body (Fig. 9.3). They fade somewhat as the force is maintained, and a

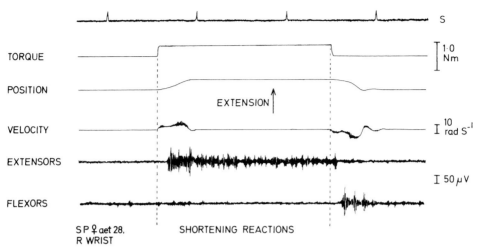

Fig. 9.4. Tonic shortening reaction in the forearm extensors, phasic shortening reactions in the flexors.

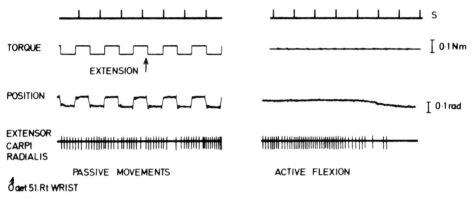

Fig. 9.5. Single motor unit not modulated by torques that cause larger movements than those associated with activity/silence depending on voluntary control. No evidence of operation of stretch reflex.

clear latency is measurable between the change in the direction of the torque and the onset of the electrical activity (Fig. 9.4).

When the person tenses very slightly, EMG activity can be recorded which is not modulated by small passive movements—while active movements of comparable amplitude are associated with large alterations (Fig. 9.5). With moderate stiffening and fairly strong forces, stretch reflexes are seen because the muscles being stretched are the ones that show electrical activity (Fig. 9.6). With maximum stiffening there is a profound change in the biomechanical properties of the arm; the motion is virtually abolished, except for a slight shudder as the torque reverses, and some tremor. Continuous EMG activity develops which is not modulated by the force (Fig. 9.7).

Thus, according to the state of the person, there may be electrical silence in the muscles, shortening reactions, stretch reflexes or continuous co-contraction.

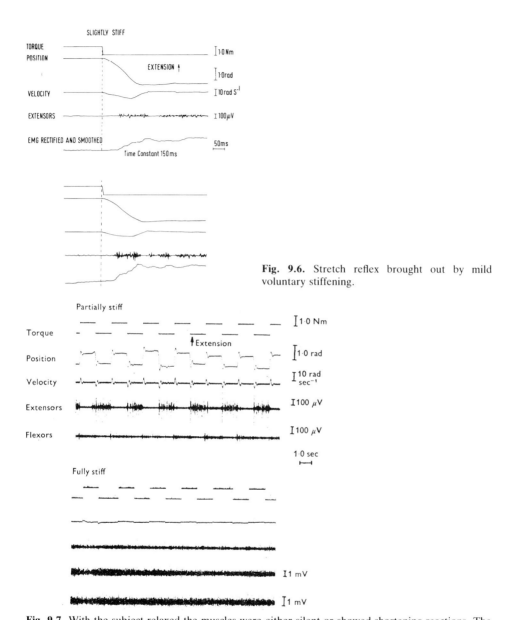

Fig. 9.6. Stretch reflex brought out by mild voluntary stiffening.

Fig. 9.7. With the subject relaxed the muscles were either silent or showed shortening reactions. The subject then stiffened the forearm, either with or without increasing the grip on the handle. With moderate stiffening to reduce the excursion to one third to one tenth of that in the relaxed state, stretch reflexes were commonly encountered. Of 12 subjects, 11 showed clear stretch activity in the extensor muscles and most of these also showed activity in the flexor muscles. The EMG activity was commonly maximal soon after the reversal of force, although substantial tonic activity was present too. The reactions were often sub-clonic, there being several oscillations at the time of the torque reversal. The latency to an unpredictable reversal of torque was about 50ms. When the person stiffened maximally there was much co-contraction and often some tremor but stretch activity was inconspicuous or absent (from Walsh 1976b).

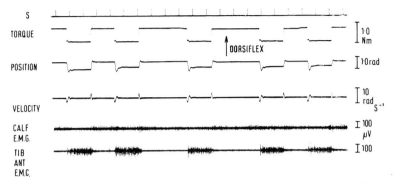

Fig. 9.8. Stretch reflexes in Holmes-Adie syndrome. Whilst people with this condition lack tendon reflexes they can stand normally, do not have any obvious problems with posture, walking or running and apart from pupillary changes are otherwise not abnormal. Observations at the ankle using rectangular alternating torques driven by a binary noise generator—transitions are thus unpredictable. Clear stretch reflexes in tibialis anterior. Male aged 21.

When an active muscle is stretched there occurs not only the short-latency response through the spinal cord (M1) but also at least one longer-latency response (M2). It is now acknowledged that the route involves loops through the cerebral cortex (Matthews 1991):

> There is widespread support for the view that cortical mediation allows for the better adjustment of the reflex to prevailing conditions. A particular instance is provided by the spread of the response to muscles other than the one being stretched; in the decerebrate cat the reflex is strictly localized, but in humans the response can spread widely with little extra delay. This was first noted in the context of posture where a variety of muscles can be activated without being directly stretched, and contribute to the maintenance of the stability of the body. Such responses depend crucially upon neural 'set', and disappear when a muscle acts in a different context ... In contrast to the spinal M1 reflex, the long-latency response can be routed to an apparent antagonist if its contraction were mechanically advantageous. Thus, stretch of the biceps muscle when it is acting to supinate the arm elicits long-latency excitation of the triceps, as well as of the biceps itself. The triceps then counteracts the unwanted component of flexion produced by the reflex activation of the biceps. Such initially unexpected synergies also occur during voluntary contraction.

Observations on a person with the Holmes-Adie syndrome

An investigation was made into the responses to low-frequency alternating torques at the ankle in a 20-year-old male medical student who had absent ankle jerks[1] and a myotonic pupil. The apparatus used is shown elsewhere (see Fig. 10.22). To randomise the timing of the alternations of torque the power amplifier supplying the motor was fed from a binary noise generator. This prevented the person anticipating the movement and the consequent possibility of interference from voluntary control. When he was partially stiff there was continuous EMG activity from the tibialis anterior electrodes when the motor was pressing in a plantarflexing direction (Fig. 9.8). This activity evidently represented a tonic stretch reflex, but no

---

[1] It is known that in this syndrome the H reflex (pp. 27–29) also is reduced in size (Krott and Jacobi 1972).

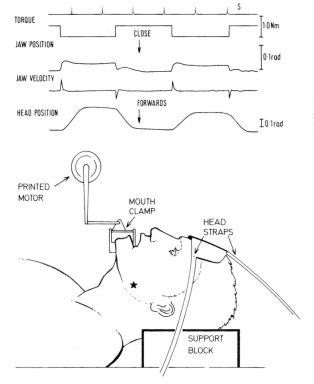

Fig. 9.9. Forces applied to jaw in sitting position, the head moves. FJW, female aged 19.

Fig. 9.10. Arrangements for investigating the biomechanics of the jaw. As the axis of rotation of the jaw changes as the jaw is opened it is not satisfactory to try to make the motor concentric. Force is transmitted from a lever attached to the motor spindle, through a link equipped with low friction pivots at each end, to the clamp on the jaw. The metal plates of the clamp are furnished with strips of balsa wood into which the teeth are sunk. The head is stabilised by the tension between the plastic support block and the metal plate on the forehead.

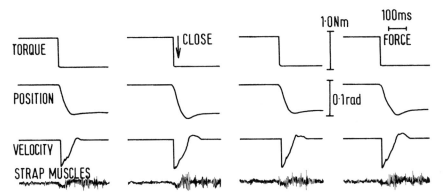

Fig. 9.11. Force changes from opening to closing direction. Peak velocity reached very rapidly. Deceleration appears to start too soon for reflex activity in the strap muscles to be initially responsible. EGW, female aged 14.

corresponding activity was found in the calf musculature. Similar responses were seen on the two sides. The author is unaware of other studies relating to the presence of tonic stretch reflexes in this condition.

Tone of the jaw musculature

The mechanical response of the jaw to an external force has been studied very little. There have been electrophysiological investigations of the jaw jerk, and Yemm and Nordstrom (1974) measured the tension needed to open the jaws of anaesthetised rats. Significant stiffness remained when the muscles were inactive. In investigating the biodynamics of the human jaw it is appropriate to apply the forces from a printed motor through a low-friction linkage because the axis of rotation varies as the position changes. The linkage was equipped with a device that clamped the lower jaw between the teeth and the underside of the chin anteriorly between the two rami of the mandible. In the initial observations the person was seated but it was found that under these circumstances there was much head movement and only a little jaw movement (Fig. 9.9). The head had to be stabilised adequately. In the arrangements adopted, the person lay with the occiput on a shaped block of firm polystyrene. A curved metal plate was applied to the forehead with three tight straps (Fig. 9.10), and the scalp was thus tensed upwards from the eyebrows. The procedure was not really uncomfortable provided the experimental sessions were reasonably brief. The motion was recorded on a conductive plastic potentiometer, concentric with the estimated position of the instantaneous centre of rotation of the jaw for the range of motion being investigated. Four young men and two young women (none of whom wore dentures) acted as subjects.

Specimen results are shown in Figure 9.11. When the torque changed from a jaw-opening to a jaw-closing direction, the jaw moved in the appropriate direction for a brief period and was then arrested. There was only slight overshoot and the velocity trace was spiky. The EMG activity of the temporalis muscle was not modulated when the force changed.

With an abruptly changing force, the jaw moves at first in the direction of the applied force but peak velocity is usually achieved in 10 to 20ms. No EMG activity could be found in the temporalis, masseter or strap muscles in the pre-braking period, and the responses seen in the strap muscles (Fig. 9.11) are too late to account for the early onset of the deceleration. The latency of the jaw jerk, about 8ms, would probably not give enough time for mechanical events in the muscle to develop significanttly.

Furthermore the latency of any such responses (if any were missed in the recordings) would probably have been longer because of the more gradual stretching of the muscles compared with the abrupt stretch when the chin is hit with a tendon hammer. The time to reach peak velocity did not lengthen as the strength of the force was reduced, although the actual velocity varied widely. This 'isochronism' would be expected if the braking was due to elasticity. The latency of a reflex should become longer when, with reduced force, the stretching to a given distance is slower.

The superficial location, the fan-like form and the relative isolation of the temporalis muscle makes it ideal for recording activity of single motor units by surface electrodes (MacDougall and Andrew 1953).

The inertia was increased in one set of observations by the use of a balanced inertia bar (0.02kg m^2). The system then became resonant, showing two or three oscillations after the sense of force reversed. The power/weight ratio at the jaw is high, a fact that may be considered in assessing the responses to added inertia. On some occasions the subject was told to attempt to stiffen the control of the jaw and so resist the changes of force. In general no changes were observed; the person did not appear to have the neurophysiological capability to stiffen (or rather to stiffen more than when 'relaxed'). The relative stiffness of the jaw is also illustrated by the pronounced head movements which resulted when the observations were made in the sitting position.

The response of the jaw to a step change of force may be considered admirable for a biological system. It is rapid and almost dead beat. The damping may possibly be due to viscosity of tonically active jaw opening and closing muscles. Reflex feedbacks are unlikely to be important, owing to the short time available for the response. In everyday life, forces will be transmitted to the jaw by motion of the body itself. As the mandible is cantilevered out from the tempero-mandibular joints there is liable to be instability of the system when subjected to vertical acceleration and deceleration resulting from running. To prevent our teeth from knocking together and breaking, a superb control system is needed. The tightness of the control, perhaps of an elastic nature adjusted by tonic co-contraction of antagonistic muscles, may provide built-in protection which can be backed up by more powerful but slower-acting reflex loops.

Reduction of muscle tone caused by J-receptor activity
J-receptors are supplied by the vagus nerve and lie in the lung. J stands for 'juxta-capillary', owing to the close proximity of these receptors to the vessels supplying the alveoli of the lung. Their physiological significance has been extensively explored by Professor A.S. Paintal and colleagues in Delhi (Paintal 1986*a*). They are sensitive to pulmonary oedema which may occur at high altitudes, under a variety of pathological conditions and when irritant gases are inhaled. Rapid breathing ensues when J-receptors are stimulated. Another effect of J-receptor activity is a feeling of generalised weakness and their effects may have seriously impaired the efficiency of Indian troops in the Himalayas during the tension with China in the 1960s. In the Bhopal gas tragedy, many people suffered from pulmonary congestion and oedema as toxic gases, principally methyl isocyanate, were released from a chemical plant. Apart from respiratory symptoms, extreme muscle weakness was observed (Paintal 1986*b*). Fleisch and Grandjean (1948) sought to investigate the effect of high altitudes on muscle tone by making observations on subjects staying in the laboratory at the summit of the Jungfraujoch, but the scope of the investigation was limited by the techniques then available.

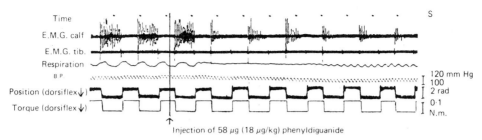

Fig. 9.12. Stretch reflex in calf of cat given sodium pentobarbitone. Stimulation of J-receptors by phenyldiguanide reduces stretch activity. There is a period of apnoea followed by rapid breathing (from Paintal and Walsh 1981).

Paintal and Walsh (1981) studied the effects of muscle tone in cats. Sodium pentobarbitone anaesthesia was used and the ankles of the animals were alternately flexed and extended by a small basket-wound motor. A compound (phenyldiguanide) known to stimulate J-receptors was injected intravenously. The injection greatly reduced the stretch reflex (Fig. 9.12). In one animal the chest was opened and the pericardial sac injected with local anaesthetic. Injection of compound into the right atrium inhibited the stretch reflex whilst injection into the left atrium was without effect, confirming that the receptors were related to the pulmonary circulation.

In further observations (undertaken by V. Virmani, A.S. Paintal and the author) a hemiplegic patient was injected intravenously with a small dose of the alkaloid lobeline. This substance was named after Matthias de Lobel (1538–1616), botanist to King James I. A dried powder of *Lobelia inflata* was smoked by the North American Indians and is known as Indian tobacco. A dried preparation is mentioned in a standard pharmacopeia (Martindale 1989).

Injections of lobeline have been widely used to measure circulation time because the carotid body is stimulated and when this occurs the patient coughs. Lobeline is known to stimulate J-receptors too (Jain *et al.* 1972). Clonus, elicited by upward pressure on the sole of the foot, was abolished by this procedure.

There are thus clear effects on the mechanisms concerned with muscle tone. Paintal developed the theory that since strenuous exercise may give rise to some lung oedema, the reflex effects of the J-receptors on muscle serve a regulatory role in terminating the muscular activity.

Smokers sometimes claim that tobacco relieves tension. This may not be purely psychological, for smoking can reduce tendon reflexes. The knee jerk is said to be depressed by smoking cigarettes to an extent that depends largely on the nicotine content. Domino and Baumgarten (1969) pointed out that:

> It may very well be that some of the reduction of skeletal muscle tension thought to occur during tobacco smoking may be on this basis.

Perhaps some of the effects of smoking are mediated through J-receptors, but the effects of smoking on muscle tone have not been systematically explored.

Fig. 9.13. Cat pressing his trunk against his owner's leg. The animal is sensing the mechanical resistance and is reacting accordingly (from Darwin 1872).

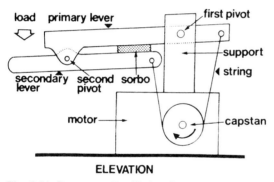

Fig. 9.14. Pugnatron. Equilibrium is reached when the force exerted by the capstan on the string attached to the primary lever is adequate to bear the weight of the system. This is attained when the string from the secondary lever is relatively slack. (From Walsh 1974, by permission.)

Pugnatron-like postural activity

A postural system does not always function merely to oppose externally applied forces. It may be 'load seeking'. In obtaining information about the external world, we use vision and hearing, but we also need 'active touch', the ability to discern the nature or the mechanical impedance of our surroundings (Fig. 9.13). Classic neurophysiology has said little or nothing about this. To obtain information about the possibilities of such a system, a mechanical model (the 'pugnatron') was constructed (Fig. 9.14).

The name 'pugnatron' was originally coined by Professor F.C. Williams FRS to describe a servo-system incorporating a positive feedback loop (Williams and

Uttley 1946). The spindle of the motor initially stood still but if twisted clockwise drove anticlockwise against the torque. When it was fitted to a toy car, the vehicle always travelled uphill.

The main lever of the device, consisting of two long plates fixed together, is supported on pillars. A pivot at one end supports a second lighter lever between the two halves of the first. The two levers are coupled through springs or rubbery material, and adjustable oil-filled pads provide damping between them. Power is supplied to the system by a smooth metal drum rotated at constant speed by an electric motor. A string is attached to the main lever and winds round the drum 2½ times. The other end is attached to the smaller lever. The drum is thus used as a capstan; it is a mechanical amplifier and the pull on the big lever is many times that provided by the small lever. The device 'measures' the mechanical resistance that it encounters and reacts in such a way as to shift the load. Thus it will move upwards under the influence of a downward-directed load such as that produced by a weight. The system would continue to move as long as the small lever was loaded if it were not for the geometrical arrangements. The 'sensing' end of the string is nearer to the main fulcrum than the 'output' end. Thus, as the arm rises, the string tends to become slack and the action of the instrument is self-limiting. The process can be regarded as depending on two systems, one of which reacts to the mechanical impedance with which it is in contact and the other of which is negative-position feedback. The effect of a load is greatest when it is at the far end of the small lever. When the load is close to the pivot the system moves down, albeit reluctantly. When the force is delivered to the lever on the other side of the pivot, a servo-assisted movement is obtained. When a force is exerted downwards on the first lever it is resisted in a manner similar to that of a stretch reflex. The device 'struggles', 'shakes hands' and will explore with to/fro movements soft material which it touches. If a brush is fixed to the sensing lever and brought into contact with a piece of cardboard, the device will 'paint' in up/down strokes until the surface is removed. To a brief tap there is a jerk-like response reminiscent of a tendon reflex. If there is little or no damping the device may oscillate in a manner reminiscent of clonus especially if the small lever is seriously unbalanced. If the 'arm' is supported the amplification of the capstan falls, for the input end of the cord goes slack and the system 'rests', leaning its weight on the support. In a variation of the model a double-acting system was devised which pushed down actively when it encountered resistance below the second lever. Photographs of the two pugnatrons have been published elsewhere (Walsh 1968a).

In making contact with the world around us we are accustomed to feeling the resistance of objects we touch, soft or hard, living or inert, reacting or passive. It is part of automatic exploratory behaviour. Load-seeking behaviour seems to be a characteristic of much motor behaviour. A quotation from Isaiah (X. 15) is perhaps appropriate:

> As if the rod should shake itself against them that lift it up.

Some years after this work a patient was tested who showed the same basic mechanisms. The clinical findings were that:

Mrs C.C., a 31-year-old white housewife, had developed tonic inversion of the right ankle at the age of 20 years during her first pregnancy. The dystonia progressed and interfered with walking, spreading first to the right upper limb and then to the left leg. For the previous two years she had been confined to a wheelchair. Apart from the inability to walk there was little impairment of other activities and she continued to run her household. She had some spinothalamic sensory impairment and imperfect bladder control. For the past six years she had been liable to powerful opisthotonic spasms during sleep and wakefulness which were liable to throw her from her wheelchair. The CSF was normal, and myelography, pneumoencephalography and electroencephalography revealed nothing amiss.

Two other members of the family suffered from an unusual neurological condition: her mother's first cousin Mr J.L. (aged 57 years) and his sister Mrs A.McL. (aged 65) had been afflicted with a slowly progressive paraplegia for eight years and 23 years respectively.

On examination intellect and cranial nerves were normal, as was the left upper limb. The other three limbs were held in tonic adduction and inversion, more so at the periphery of the limbs. There was a consequent reduction in the facility of voluntary movement. She had normal position and vibration sense but pain and temperature were impaired below the fourth cervical dermatome on the right side and the fourth dorsal on the left.

Tendon jerks were brisk and symmetrical, Hoffmann's sign was positive and the plantar reflexes extensor, but both these abnormal reactions exhibited a curious delay. The disorder of tone disappeared during sleep and general anaesthesia.

A stereotaxic operation on the left side reduced the rigidity of the limbs on the right side of the body. Three months later the observations recorded below were made.

Taken as a whole the observations were consistent with an organic disorder of the brain but the nature, extent and position of the lesion or lesions were not ascertained. The observations were made in 1973 before brain scans were available.

When the foot was pushed down by hand, it felt as if the displacement was being actively resisted because extra force was being brought in by the dorsiflexors. The feel was quite different from that of a typical rigid or spastic limb, and the unusual nature of the response was verified by instrumental observations. The foot was tested using apparatus of the type shown later (see Fig. 10.22). When a steady plantarflexing force was switched on, the initial movement in that direction was rapidly checked and within a second or so the old position had been re-established; with the further passage of time dorsiflexion continued, a plateau being reached after several seconds (Fig. 9.15). It was clear that the applied torque had been more than matched by a force produced in the muscles of the leg; the effect may perhaps be described as 'action and super-reaction'. Throughout the procedure the EMGs

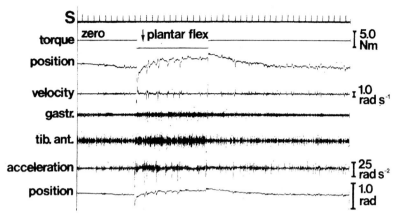

Fig. 9.15. Motion of ankle in response to steady plantarflexing force. Final position is one of dorsiflexion. The position changes have been recorded at two different sensitivities. (From Walsh 1974, by permission.)

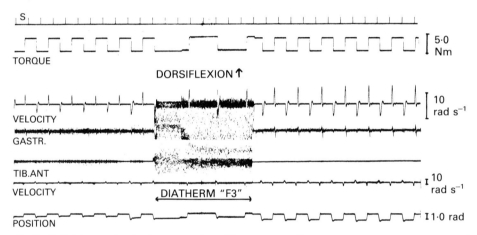

Fig. 9.16. Pulvinar lesion in dystonic patient. The pugnatron-like responses are not seen because the torque was alternating from dorsiflexion to plantarflexion and not from zero to plantar flexion. Because of the requirements of the arrangements in the operating theatre, it being necessary for the patient to be face downwards on the operating table, the axis of the ankle was horizontal, the foot moving vertically. The stiffness was reduced immediately after the pulvinar lesion; the timing is given by the interference from the diathermy apparatus.

from both channels, but especially that from the tibialis anterior, showed increased activity. When the force was withdrawn, the foot sprang further into dorsiflexion and then gradually returned towards its original position. The responses must have been reflex as there was no voluntary power in the foot. One can only speculate about neurophysiological mechanisms. Animal work has at times provided evidence that the muscle spindles may reflexly excite gamma motorneurons. There may be heightened gain in this circuit, which is normally almost quiescent, causing positive feedback.

There were reasons for feeling that small lesions of the pulvinar might be useful in treating muscular hypertonia, and that the possible complications following a bilateral pallido-thalamotomy might thus be avoidable. The neuro-anatomical basis for this surgical intervention is however unclear. Data on 11 patients subjected to the operation are given by Gillingham *et al.* (1977).

A second stereotaxic operation was undertaken ablating part of the pulvinar. The rigidity of the left foot was relieved and the pugnatron-like activity abolished temporarily (Fig. 9.16).

Later a 13-year-old boy, also with dystonia, and with the same curious feel on manipulation, was tested and found to behave similarly. The phenomena described evidently correspond to that called '*gegenhalten*' by Kleist (1927).

Conclusion

The author trusts that it will now be clear to the reader that the use of these alternating torques gives information which cannot be obtained by the use of sinusoidal forces alone; the various procedures are complementary, not alternatives.

10
HYPERTONIA: RIGIDITY, SPASTICITY AND OTHER CONDITIONS

Physicians and surgeons now have a wide choice of treatments for some forms of hypertonia, and much of the information in this chapter was obtained while assessing the efficacy of different procedures. With the introduction of L-DOPA my clinical colleagues were interested in machine measurements to evaluate the efficacy of the drug and find appropriate dosages. More than 200 patients with Parkinsonism were tested accordingly. Some of the observations on hemiplegic patients were made during an evaluation of the efficacy of inflatable splints, a procedure sometimes used in attempts to reduce spasticity[1].

Parkinsonism[2]
Observations during operations
Some of the earlier observations were undertaken during stereotaxic operations at the Western General Hospital. Burr holes had been made in the skull a few days before, but no general anaesthetic was used during the stereotaxic procedure and the patient was conscious. He was face downwards. One hand was coupled to a printed motor. Before these investigations the effect on rigidity was assessed during the operation by one of the surgeon's assistants manipulating the wrist. Changes in resonant frequency during the operation were recorded in eight cases. A fall was seen as the needle was initially passed with further lowering following coagulation, by diathermy, in the ventrolateral nucleus of the thalamus and globus pallidus.

With experience, measurements of resonant frequency came not to be regarded as the method of choice for measuring rigidity, for it was important to give the surgeon information about progress as soon as possible. Sometimes low-frequency alternating torques were used. An example of a recording where a low-frequency triangular torque was used during a thalamo-pallidotomy is shown in Figure 10.1. This operation has been superseded by medical treatments, so it is unlikely that further data will be obtainable.

Resonant frequencies
Patients undergoing this treatment were assessed preoperatively and postoperatively. The changes in resonant frequency are shown in Figure 10.2. Data from

---

[1] Some reduction of tone was observed in a minority of 14 patients tested but the results were variable and have accordingly not been the subject of a publication.
[2] James Parkinson (1755–1824), London physician, did not mention rigidity in his description of the 'shaking palsy'—perhaps because, as was customary in that era, he did not examine his patients (Parkinson 1817).

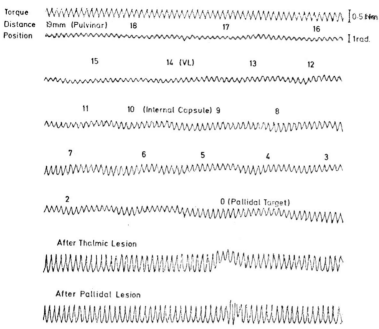

Fig 10.1. Flexibility of right wrist during left thalamo-pallidal stereotaxy. Continuous recording—with triangular wave torque at 1Hz shown on top line. The other traces represent position. The passive movement increases as the electrode passes through the ventrolateral thalamus (VL), internal capsule and finally pallidum. After the thalamic and pallidal lesions the wrist is very flexible. Numbers above trace correspond to mm posterior to anterior commissure (target in globus pallidus). Extension upwards.

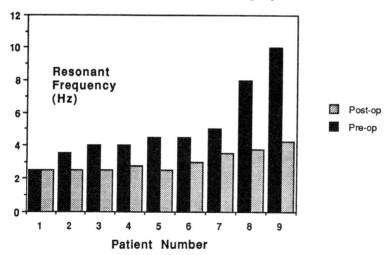

Fig. 10.2. Parkinsonism: resonant frequencies of wrist before and after stereotaxic intervention. (Data from Gillingham *et al.* 1973.)

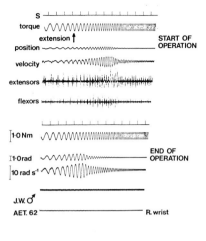

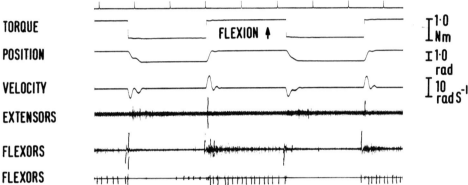

Fig. 10.3. Chirps: drop of resonant frequency in Parkinsonism as a result of stereotaxic intervention. The EMG becomes silent.

Fig. 10.4. Shortening reactions. The lowest trace was recorded using fine wires inserted into the muscle and shows a large single unit. Male aged 55, post-encephalitic Parkinsonism.

patients with continuous tremor have been excluded. With intermittent tremor the measurements refer to periods when the tremor was quiescent. The reductions in resonant frequency were greatest when it was raised at the start, but for the most part they remained somewhat elevated after the procedure. A recording of the changes during one of the operations is shown in Figure 10.3.

Low-frequency abruptly alternating torques

When a force was applied to the wrist in this way, three main types of response were encountered.

1. SHORTENING REACTIONS (Fig. 10.4)

These are generally normal. They appeared to be particularly conspicuous in the patients with Parkinsonism. Westphal (1880) described a 'paradoxical reaction'. Denny-Brown (1962) gave this account of the phenomenon:

> If the plantar flexor of the foot or palmar flexor of the wrist is suddenly stretched, the response in

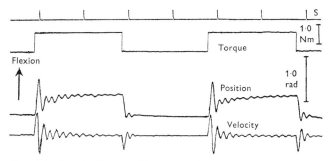

Fig. 10.5. Cogwheeling. Oscillatory transients at 6.5Hz, more prominent for extension. Spontaneous tremor at 5Hz at other times. Female aged 46, Parkinsonism for seven years. Left wrist pronated, *vertical* movements (from Walsh 1970*b*).

> the shortened antagonistic extensor is often visible to the eye as a brief contraction of the tendons of the corresponding pretibial or wrist dorsiflexor muscle.

This response would appear to be an exaggeration of a normal shortening reaction.

2. COGWHEELING (Fig. 10.5)
An early description of this sign is that of Moyer (1911):

> It is elicited by the examiner grasping the wrist with one hand and steadying the arm with the other, above the elbow. Rapid flexion and extension of the arm is made. Instead of an even movement, without resistance, when there is no involvement of motility, one, two, or, perhaps, three slight hindrances to the movement are experienced by the examiner, which communicates to the hands of the examiner a jerky feeling . . . After the extremity has been passively moved for a short time, the jerks will slowly disappear, to be followed by a return after a period of rest.

When the force changes at the wrist abruptly into flexion the hand moves in the direction of the applied force and then brakes in a manner that is strikingly discontinuous. The velocity trace shows a series of decrementing oscillations characteristic of a badly damped system. The waves are faster when the limb is held stiffly, and range between 3 and 8Hz. The first swing is sometimes large enough to bring the hand momentarily back to the starting position.

The cogwheel phenomenon is often regarded as masked tremor. An autonomous central clock responsible for spontaneous tremor (pp. 140–142), however, is not involved for two reasons: (i) the initial swings are phase-locked to the step function, and (ii) while the rate of Parkinsonian tremor is not changed by adding inertia the ringing of the cogwheel effect is slowed. An addition of 0.01 kg m^2 in one patient changed the rate of the transients from 6 to 2½Hz. It would appear that patients with Parkinsonism have two forms of instability with rather similar frequencies.

Cogwheel rigidity may become apparent only when the person partially stiffens the limb (Fig. 10.6). It is an early sign of the disease and often disappears with progression of the malady. It would appear to be due to active phasic stretch reflexes—without the 'follow up' of a tonic component to hold the posture. Tonic stretch reflexes are not prominent features of most patients with Parkinsonism.

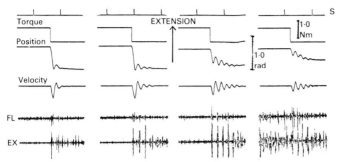

Fig. 10.6. Cogwheeling. Oscillatory transients occur when the force changes into flexion. They occur when the patient stiffens a little. Rhythmic EMG discharges. Female aged 48.

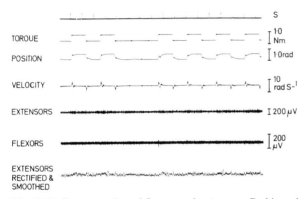

Fig. 10.7. Co-contraction of flexors and extensors. Parkinsonism, 10-year history. The position trace shows the slow movement characteristic of creep. Male aged 65, extension downwards.

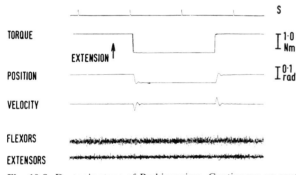

Fig. 10.8. Dystonic stage of Parkinsonism. Continuous co-contraction of antagonistic muscle groups. Very little motion in response to change of torque and no stretch reflexes. Male aged 62.

3. UNMODULATED CO-CONTRACTION (Figs 10.7, 10.8)

In a late stage of Parkinsonism the limbs may become very stiff with co-contraction of flexors and extensors. When the force changes there is no alteration of the EMG activity. The movement for a given force is slight and the new position attained rapidly. The velocity trace is accordingly spiky.

Two types of activation

Rigidity and tremor in Parkinsonism commonly fluctuate both in an apparently spontaneous manner and in response to identifiable mental and physical activity. One clinical procedure is to get the patient to subtract seven from 100 successively while manipulating the arm to feel for changes in resistance.

Webster (1966) used an isokinetic method of measuring the work done in moving the elbow of Parkinsonian patients (p. 38). He pointed to the need to use a form of activation:

> First, it was noted that as the average patient with Parkinson's disease became drowsy, rigidity values fell to almost zero, indicating that central phenomena were involved. Second, an opposite effect was found, namely, a tendency for rigidity to be precipitated by an alerting reaction. Because both of these effects are powerful, we feel little reliable data can be obtained without an effort to control the activation level. Therefore, our daily rigidity measurements now include 10 passive motion cycles at rest, all averaged into a single resting value, and 15 cycles of rigidity measured under activated conditions. The activation phenomenon is produced by having the patient operate an electronic pursuit meter with one arm while rigidity is recorded from the opposite arm. We thus get not only an alerting phenomenon for the rigidity measurement but also a pursuit score indicating performance.

The alterations resulting from a variety of methods of activation have been studied with special emphasis on the use of a telephone and morse key. Patients with severe continuous tremor and high levels of initial rigidity are generally little influenced by activation, the most striking results are found in moderate and early cases, in whom the tone at rest may be normal. Forty-seven affected wrists have been investigated, and all gradations of the condition have been included.

The patients were asked to remember a telephone number and to dial it with the hand that was not coupled to the motor. This procedure does not influence tone in normal people. Increases in tone were observed on 29 occasions; the changes were often substantial (Fig. 10.9). The compliance of nine normal people at rest was 2.70 ± 0.56 rad/N m (mean and SD) and when activated the value was essentially the same, 2.64 ± 0.78 rad/N m. In eight Parkinsonian patients who had not been subjected to surgery the resting compliance was 1.82 ± 0.86 rad/N m; this fell to 0.91 ± 0.51 rad/N m when using the telephone. There had been a stereotaxic operation in 13 patients; in 11 of these, activation by the use of the telephone did not occur. Dialling was a potent activator of rigidity in the unoperated cases.

Telephoning is poor at provoking tremor (three times only, inhibition once). However, tapping a morse key as rapidly as possible with the opposite hand, or a switch with one of the feet, elicited tremor in all of the 16 susceptible patients. This was not due to the passive transmission of vibration, for on the whole the frequencies were different. This procedure activated rigidity 29 times but the interpretation of stiffness in the presence of powerful tremor is complicated. The two types of activation clearly operate through different mechanisms.

Tremor

The velocity trace shows up irregularities of motion more conspicuously than the

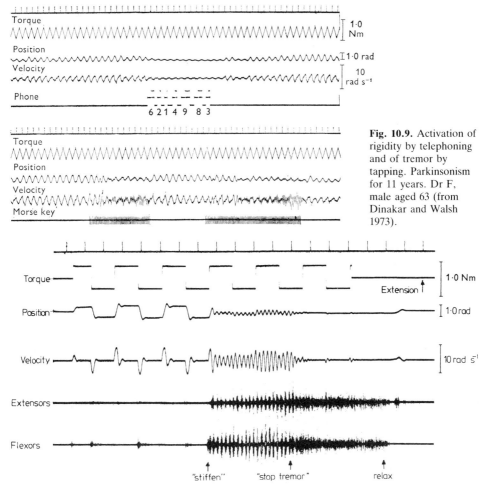

Fig. 10.9. Activation of rigidity by telephoning and of tremor by tapping. Parkinsonism for 11 years. Dr F, male aged 63 (from Dinakar and Walsh 1973).

Fig. 10.10. Low-frequency alternating torques. Voluntary stiffening of wrist brings out latent Parkinsonian tremor and the movements in response to the applied force are greatly reduced. When told to stop he could do so, and the rhythmic EMG discharges were replaced with continuous co-contraction. When told to relax the co-contraction was abolished. Male aged 59 (from Walsh 1979).

position trace. In many patients the tremor of wrist movement is seen to be remarkably sinusoidal; this is true even when the velocity trace is examined. The tremor shows considerable slow variations of amplitude but the frequency is often very stable. Even when records are taken several months apart, the rate is usually unchanged. Very occasionally there may be some increase of rate with voluntary stiffening or with sustained flexion and extension. Overt tremor may come and go and even when absent it is sometimes possible to detect EMG activity at the tremor frequency. On such occasions it may be possible to see a flickering movement of the skin over a part of the muscle which is evidently not nearly powerful enough to move the hand. The forearm flexors and extensors are in general reciprocally active but with the hand held actively in extension or flexion the rhythmic action may be confined to the extensor or flexors respectively.

Tremor may be suppressed when the forearm muscles are tightened voluntarily, but in other subjects the manoeuvre may bring out a latent tremor (Fig. 10.10).

Fifty-five patients were tested for movements in the horizontal plane. All were cases of idiopathic Parkinsonism except for one where the disease followed exposure to carbon monoxide and another where it followed an attack of encephalitis lethargica many years previously. The rate of the tremor on the right was 5.25 ± 0.78Hz (mean and SD), and that on the left 4.48 ± 0.54Hz. The difference is statistically significant and unexplained[3].

Beats were seen when rhythmic torques were applied to the wrist at a frequency close to that of the tremor (Fig. 10.11). There was no tendency to entrainement. Beats could similarly be seen when observations were made at the ankle.

The study of the effects of rhythmic inputs on oscillators has a long history (see review by Walsh 1979). Several renowned scientists have been interested in this problem. Huygens[4] noted that two clocks mounted on the same board came into step. Rayleigh[5] found that two electrical tuning forks synchronised when fed from the same battery, and observed the same phenomenon in two organ pipes fed from the same wind chest. In these examples the transfer of tiny amounts of energy led to entrainement, one oscillator influencing another. Appleton[6] made a study of the mathematics underlying the synchronisation of a valve oscillator by a powerful distant transmitter. Synchronisation depends on non-linearity. All real oscillators are non-linear, but this can take an unlimited number of forms. Firm conclusions about the significance of the production of beats can only come if we find out more about the non-linearities. The mathematics are highly involved.

Two theories of the origin of tremor and clonus are represented in Figure 10.12. The stability in frequency under a wide variety of peripheral conditions reinforces the view that the tremor of Parkinsonism results from the activity of a central generator or generators; it is evidently not 'servo-driven'. The EMG discharges were never increased when the force of the motor was opposing the tremor and producing a minimum in the beat envelope. There was thus nothing to suggest that the motion was 'servo-assisted'.

Tremor is sometimes found widely in the body. When records were taken from different sites there appeared small but definite differences of frequency

---

[3]The CNS is asymmetrical; *e.g.* 80 per cent of newborn babies turn their heads to the right for preference.
[4]Christiaan Huygens (1629–1693), Dutch mathematician, astronomer and physicist.
[5]Lord Rayleigh (1842–1919) made fundamental discoveries in acoustics, and showed that the blueness of the sky was due to scattering of light by small particles in the atmosphere. He discovered the rare gas argon and was awarded the Nobel prize.
[6]Sir Edward Appleton (1892–1965), physicist. He discovered the 'Appleton' layer in the ionosphere and was awarded the Nobel prize. He was Principal of Edinburgh University (1949–1965).

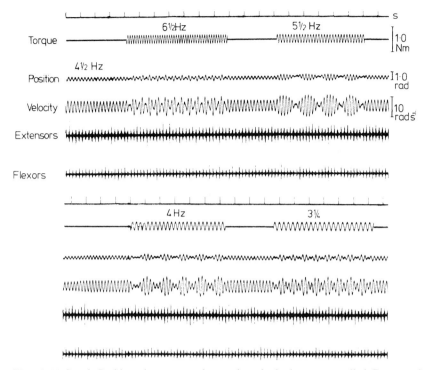

Fig. 10.11. Steady Parkinsonian tremor—beats when rhythmic torque applied. Beat envelope depends on difference between frequencies of applied torque and tremor rate. The EMG is not modulated—no evidence of stretch reflex activity. Patient female aged 67 (from Walsh 1979).

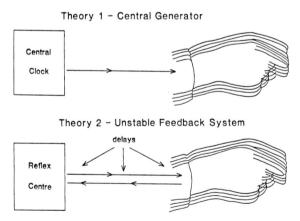

Fig. 10.12. No single mechanism covers all different types of tremor, each must be considered on its merits. There are two main theories of the genesis of pathological tremors and clonus. First, there may be a central clock, or tremorogenic zone somewhere in the central nervous system. Second, the tremor may result from an unstable feedback system. Peripheral perturbations may change the rhythm when the underlying basis is a feedback loop involving the limb.

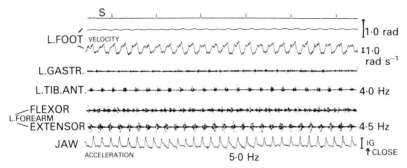

Fig. 10.13. Recording of tremor of left foot and left forearm together with a record obtained by the use of an accelerometer on the jaw. The rates have small but definite differences, the tremor in different regions is not synchronised.

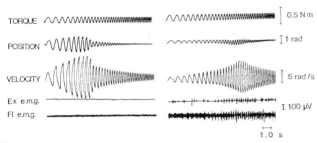

Fig. 10.14. The normal wrist compared with the spastic. The spastic *(right)* has a considerably higher resonant frequency and there is modulated EMG activity which is absent on the normal side (from Lakie *et al.* 1988).

(Fig. 10.13). If indeed the tremor had been synchronised throughout the musculature there would have been a widespread shudder at each beat, but this was not observed. It would appear that there is not one single generator but several or many driving different regions.

Spasticity
Resonance at the wrist in hemiplegia
Thirteen children aged between five and 15 were tested, and some investigations were repeated over a period of five years (Brown *et al.* 1987). None of the patients were in the acute phase; all were on a clinical plateau[7]. It rapidly became apparent that, as expected, the resonant frequencies of spastic hemiplegic limbs are elevated. An example of the use of chirps for this investigation is shown in Figure 10.14.

The data for the children are shown in Figure 10.15. The resonant frequency on the hemiplegic sides was 4.9 ± 2.3Hz (mean and SD), that on the normal side 2.3 ± 0.4 Hz. As the age of the children varied from five to 15, different levels of

---

[7]Abnormalities on the 'normal side' in hemiplegia can often be demonstrated by the quite vigorous movements by which stretch reflexes are activated in normal people (Thilmann *et al.* 1990). Comparison with data from age-matched unaffected people revealed no differences in resonant frequencies.

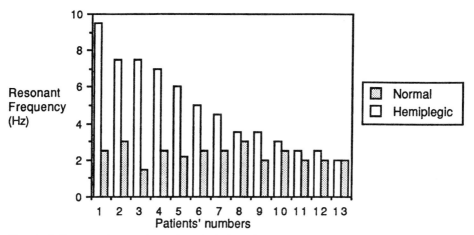

Fig. 10.15. Resonant frequencies at the wrist in juvenile hemiplegia, 13 children were tested. The data are presented in tabular form in Lakie *et al.* (1988).

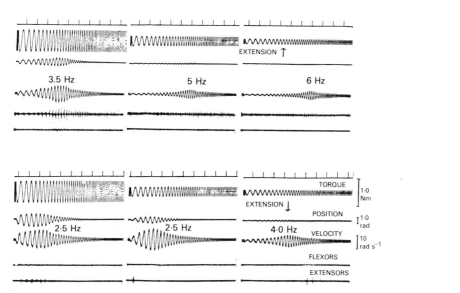

Fig. 10.16. Resonance at the wrist at three torque levels. *Top:* spastic (right) side. *Bottom:* normal (left) side. At each setting of torque the resonant frequency is higher on the spastic side. On the spastic side there is EMG activity in the flexor channel. Male aged 47—injury dating from a wartime glider crash 30 years previously.

torque were used for different patients. The torque values used were the same on the two sides and were adequate to bring the frequencies onto the flat part of the curve relating torque and frequency. The mean values show that the average frequency on the spastic side had doubled; the stiffness was therefore rather more than four-fold.

The results of testing a long-standing hemiplegic at three levels of torque are

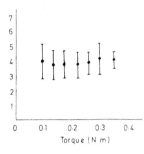

Fig. 10.17. Resonant frequency at the wrist *vs* applied torque in 21 adult hemiplegics. There was no variation within the limits tested, the rise at the lower levels of torque that occurs in normal limits was not seen. The values are however high (from Lakie *et al.* 1988).

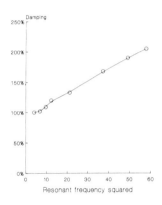

Fig. 10.18. If the damping was constant the peak velocity would have remained the same with varying degrees of stiffness. In this graph the velocity at resonance on the normal side has been divided by that on the hemiplegic side. The graph represents the results from eight hemiplegic children. It is clear that damping increases with stiffness (from Lakie *et al.* 1988).

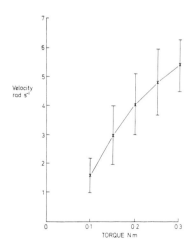

Fig. 10.19. In a perfectly linear system there would be a straight-line relationship between torque and peak velocity at resonance. The results with nine adult hemiplegics are shown. The relationship is evidently quasilinear (from Lakie *et al.* 1988).

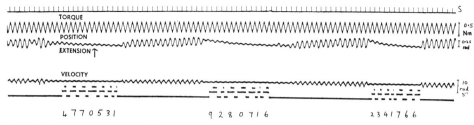

Fig. 10.20. Activation of the right wrist in spasticity due to hemiplegia by telephoning.

shown in Figure 10.16. Twenty-one other adult hemiplegics were tested at seven levels of peak torque (Fig. 10.17).

The damping was investigated by dividing the velocity at resonance on the normal side with that on the hemiplegic side at the same peak torque level. The results with eight children are shown in Figure 10.18. The damping increased with increasing resonant frequency. It is possible that this is due to the phasic stretch reflexes seen in the EMG, for these will provide negative velocity feedback. This change offsets the effects of increasing stiffness (equation 8, p. 49), the records confirmed that the sharpness of tuning was little different from that of a normal relaxed wrist.

In a perfectly linear system with viscous damping there would be a linear relationship between torque and peak velocity at resonance (pp. 57–58). The results with nine adult hemiplegics are shown in Figure 10.19. There is perhaps some deviation from linearity. Such non-linear damping does not significantly compromise the analysis of the system based on equation 8 (p. 49). According to Burton (1968):

> Even when the damping is known *not* to be viscous damping, approximating it as such often leads to a good practical estimate of the behavior of the system.

A limited number of observations were made on the effects of activation on muscle tone. An example in which the right wrist was subjected to a slow triangular torque while dialling a telephone number with the left hand is shown in Figure 10.20. The motion is decreased during the procedure. Webster (1964) compared the effects of activation in Parkinsonism with those in spasticity. He used a procedure in which the knee was subjected to stretching at 20° or 40°/s to and fro while the person undertook a pursuit task (see p. 38). The effects increased tone much more clearly in the patients with Parkinsonism.

Decerebrate rigidity
The instrument was used for measuring the resonant frequency at the wrist in patients who were decerebrate as a result of head injury. The values for seven severely rigid patients were 9.1 ± 1.5Hz (mean ± SD). There were periods of spontaneous relaxation when the values were 2.9 ± 0.56Hz. There was a further small fall after treatment with relaxant drugs (Tsementzis *et al.* 1980). The torque used for different patients was between 0.32 and 0.41N m.

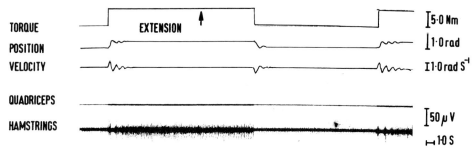

Fig. 10.21. Stiff left knee—cervical spondylosis—stretch reflex in hamstrings. Male aged 66.

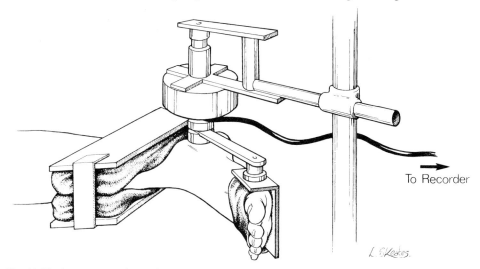

Fig. 10.22. Arrangements for testing the ankle. Shown is a G9M4 motor suitable for children. For the adults a larger motor such as the G16M4 is more appropriate. In some observations the motor was under rather than over the ankle.

Cervical spondylosis—stiff leg

Figure 10.21 shows the record of a man whose main complaint was that of a stiff leg. When coupled to a motor for testing the knee (see Fig. 8.4), it was found that a tonic stretch reflex was present in the hamstrings.

Ankle clonus

Clonus is a common neurological sign and one which is at times disturbing for the patient. The usual view of its genesis is that it is the result of hyperactive stretch reflexes re-exciting themselves (Creed *et al.* 1932), mainly because it is commonly associated with brisk tendon jerks, which in a spastic man or decerebrate cat may merge imperceptibly into clonus. Clonus may occasionally occur, however, when tendon jerks cannot be obtained as during general or spinal anaesthesia. The view was also supported by experiments on the silent period of the EMG which follows the synchronised discharge associated with a tendon jerk. These must now be

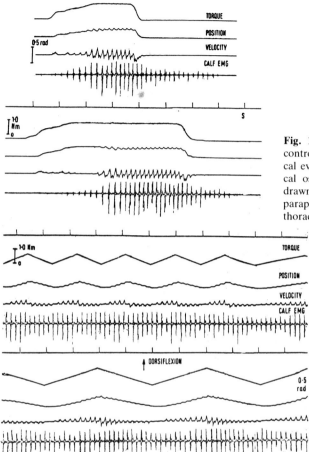

Fig. 10.23. Dorsiflexing bias, manually controlled. Two examples in both electrical evidence of clonus outlasts mechanical oscillations when the bias is withdrawn. JB, male aged 23, champion paraplegic weightlifter, complete midthoracic lesion (from Walsh 1976c).

Fig. 10.24. Slow variation of bias, dorsiflexing torque varies in triangular manner. Size of clonus and shape of pulsations vary but the frequency is unchanged. JB, male aged 23, champion paraplegic weightlifter, complete midthoracic lesion (from Walsh 1976c).

regarded as purely of historical interest. It was accordingly felt that a re-evaluation of the available evidence was long overdue.

Nine patients aged between 11 and 55 years were investigated after sustaining ankle clonus as a result of spinal injuries or disease. The procedures were not disturbing, and in fact one patient fell asleep during the investigations. The apparatus is shown in Figure 10.22. In all the patients a dorsiflexing force was needed to bring out the clonus which was at a rate of 5.5 to 8 Hz. If the bias is applied gradually, the clonus may take some time to build up.

Clonus is more jerky than the tremor of Parkinsonism or a rhythmic voluntary act; the discharges occur with almost machine-like regularity. It is difficult to see it as a 'release' of normal neural activity by the withdrawal of inhibition. The mechanical oscillations disappear if the biasing force is withdrawn, but rhythmic bursts are seen in the EMG for a variable period thereafter (Fig. 10.23). The

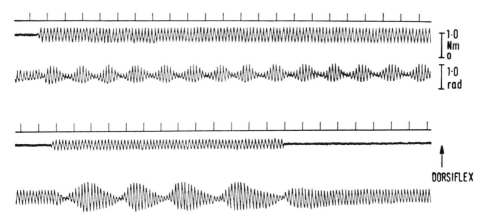

Fig. 10.25. Interference between clonus at 6Hz and rhythmic force at 5.5Hz *(upper)*, and at 5Hz *(lower)*. Beat frequencies of 0.5Hz and 1Hz respectively correspond to the difference in rate between clonus and applied torque. Male aged 23 with traumatic paraplegia of four years duration (from Walsh 1971).

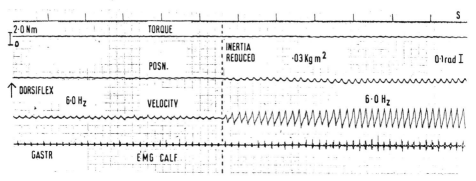

Fig. 10.26. Dorsiflexing bias—clonus. In the left half of the figure there is added inertia. When this is removed the clonus becomes more vigorous but the frequency remains the same.

waveform of the clonus varied with the bias although the frequency remained constant (Fig. 10.24).

When, in addition to the dorsiflexing bias, a rhythmic force was added at a frequency not far removed from the spontaneous oscillations, there was waxing and waning of the envelope which clearly represented beats (Fig. 10.25). When inertia was added to the system the amplitude of the oscillations was reduced but their frequency remained unchanged (Fig. 10.26).

With a dorsiflexing torque applied to the ankles of spastic patients there were often a few oscillations. If inertia was added, these non-sustained oscillations were slowed. If sustained clonus was established following the initial swings, only the first few swings were slowed by added inertia. My theory is that soon after the initial swings (which may be modified or driven by reflexes), proprioceptive discharges elicited by the tonic stretch of the calf musculature, 'switch on' a central generator which then operates at its own rate.

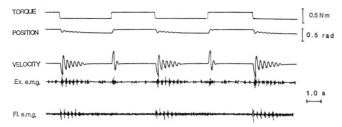

Fig. 10.27. Juvenile hemiplegia. The wrist is alternately pushed into flexion and into extension when the musculature becomes rhythmically active for a number of oscillations. Subclonic beats. Boy aged seven years with left-sided hemiplegia.

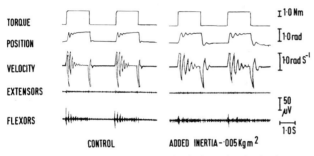

Fig. 10.28. Sub-clonic beats due to rhythmic action in the forearm flexors when the torque switches to extension. The rate of the oscillations is slowed by adding inertia. Female aged 42 with right hemiplegia.

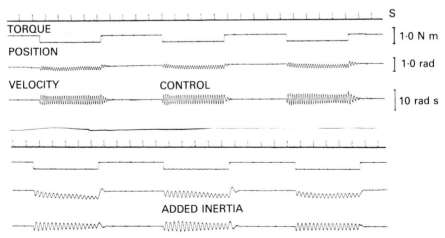

Fig. 10.29. Wrist clonus. Bias alternates abruptly between 0 and 0.25N m to dorsiflex the wrist. During dorsiflexion a regular clonus develops. Added inertia (0.002kg m^2) slowed the clonus from 6 to 3.25Hz. Male aged 39, with cervical spondylosis and clonus also of both ankles.

Wrist clonus
In the investigation of the hemiplegic children and adults, low-frequency abruptly alternating rectangular torques were used in addition to the chirps described above. With the hand on the instrument (so that there was minimal friction and its weight was supported), the wrist usually took up a flexed position. The forearm flexors showed tonic EMG activity; activity in the extensors with the hand at rest was exceptional.

When the torque changed from extensor to flexor, there were normally some oscillatory transients and several related rhythmic discharges in the flexor EMG channel (Fig. 10.27). The transients are slowed by added inertia (Fig. 10.28). These responses clearly point to a phasic stretch reflex and may correspond to the clasp-knife reflexes of classic neurophysiological texts.

Sometimes the oscillations did not die away but a sustained clonus developed as long as the force was acting in the flexor direction. Six patients who showed such sustained clonus were studied (Fig. 10.29). The addition of an inertia bar very clearly slows the oscillations, unlike in the case of ankle clonus, which supports the classic theory that the clonus is the result of self re-exciting stretch reflexes. For wrist clonus, too, beats could readily be seen on the application of an appropriate rhythmic force.

Patellar clonus
Two hemiplegic patients with patellar clonus were studied while lying supine on a couch with their knees straight. Steady bias provided by a padded bar, moving in the horizontal plane and connected to a printed motor, stretched the quadriceps and initiated the clonus. When a rhythmic component was added to the bias, entrainement of the motion was never seen even when the two frequencies were within 0.25Hz of one another and the force of the motor was about equal to that generated by the rhythmic action of the muscles. On the contrary, beats were established (Fig. 10.30). It is surely unusual to come across a system that moves more vigorously when its load is increased: yet when inertia was added to the system the oscillations increased in size. As there was no change of frequency, it would seem that a central generator was responsible for the clonus and the added mass tuned the peripheral structures so that a given drive was more effective.

Pes equinus[8]—serial casting
Little's disease[9], club foot or pes equinus, is a common deformity of cerebral-palsied children. The foot is plantarflexed and this interferes with walking since only the ball of the foot can touch the ground. The altered position may be the result of excessive plantarflexion of the ankle, flexion of the joints in the foot itself (forefoot equinus) or a combination of the two factors. The toes are often

---

[8]The term pes equinus literally means 'foot of a horse'; the hoof corresponds to a toe.
[9]William John Little (1810–1894), who suffered from the condition himself, was the pioneer of orthopaedic surgery in England. Lord Byron and Sir Walter Scott also had clubbed feet.

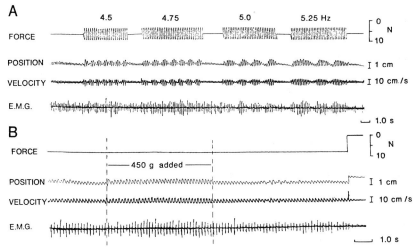

Fig. 10.30. *A:* Relationship between force and motion of the patella. A steady bias stretches the quadriceps throughout and on four occasions a rhythmic component of force is also introduced at the indicated frequencies. Beats are set up at rates corresponding to the difference in frequency between the applied force and that of the clonus. *B:* The effect of adding mass to the system is to *increase* the amplitude of the oscillation. When the bias is withdrawn the clonus peters out. Female aged 47 with left hemiplegia (from Walsh and Wright 1987*b*).

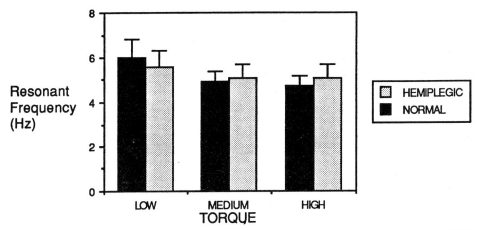

Fig. 10.31. Resonance at the ankle in juvenile hemiplegia. The three torque levels used were 0.210, 0.315 and 0.420N m. At the highest torque level the difference between the two sides is significant (p = 0.02) but not at the other values. The data have been presented in tabular form in Walsh *et al.* (1990*a*).

dorsiflexed—an adaptation, apparently, to the necessity for the toes to clear the ground.

It was accordingly decided to obtain information about the stiffness of the ankle in cerebral palsy. Fifteen juvenile hemiplegics were tested. The resonant frequencies are given in Figure 10.31. The values on the two sides are rather similar; at the higher torque value the difference is significant but quite small. In

TABLE 10.I

Electric goniometry (measurements expressed as degrees from right angle)

	Control	After first cast	p	After second cast	p	Normal feet (N 4)
Resting position	38.9 ± 11.9	32.4 ± 6.5	0.08	25.9 ± 5	0.004	24.8 ± 14.4
Passive dorsiflexion	14 ± 5	18.4 ± 9.7	0.25	18.7 ± 10.4	0.17	29.5 ± 19.1
Active dorsiflexion	10.4 ± 11.2	20.1 ± 16.9	0.09	10.4 ± 9	0.48	25.9 ± 14
Passive plantarflexion	56.9 ± 18.4	60.5 ± 13	0.33	59.8 ± 15.8	0.38	58.3 ± 20.5
Active plantarflexion	47.1 ± 9	47.5 ± 8.6	0.46	47.5 ± 20.9	0.49	56.1 ± 11.5

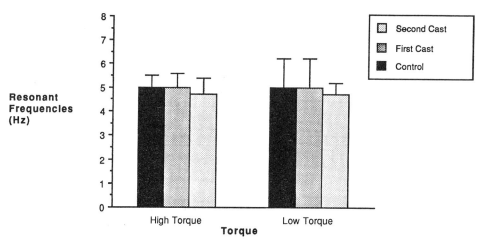

Fig. 10.32. Resonant frequencies at the ankle of children being treated by serial casting. The peak-to-peak torques used were 0.42 and 0.21N m. The data have been presented in tabular form in Walsh *et al.* (1990*a*).

only three of the 15 children was there active EMG activity modulated by the rhythmic movements.

Another group of children with pes equinus, two with spastic diplegia and five with juvenile hemiplegia, were tested. There were thus nine feet with this deformity. A cast was applied for two weeks and was then replaced with a second cast providing further stretching for two more weeks—'serial' or 'Bay' casting. This is a therapy commonly used for this condition.

When the motor was switched off it was possible to use the output of the potentiometer to record joint angles (Table 10.I). There was thus a highly significant alteration in the resting position of the foot which approached normal values. The other parameters did not change. Resonant frequencies at the ankle were also measured (Fig. 10.32). While at the lower torque level the change after the second cast did reach significance, the alterations were all trivial. The resonant frequency of four normal feet at the lower torque level was 5.1 ± 0.25Hz (mean and SD) and was apparently higher (p = 0.04) than that of the treated feet after the second cast. At the higher torque level the resonant frequency of the normal feet was 4.7 ± 0.6Hz, essentially the same as that of the treated feet (p = 0.5).

The resonant frequencies changed only marginally, and the one significant alteration—at the lower torque level after the second cast—can probably be explained by the wasting which inevitably follows the immobilisation of a limb in a cast. Before treatment, the resonant frequencies were little or no higher than the values for the small number of normal feet tested. The resonant frequency of the hemiplegic hand is often very clearly elevated compared with that of a normal limb (see above).

The physiological basis of pes equinus thus appears to be quite different from that of the hemiplegic wrist; so probably there is not increased stiffness but rather the muscle or tendon, or both, are shorter. This finding has significant implications for the appropriate therapy. The changes have undoubtedly arisen because of abnormal innervation at some time previously, presumably in infancy. Anatomical changes have occurred which persist when no neurally induced stiffness can be detected. Correctable equinus due to increased tone progresses to fixed equinus due to muscle shortening.

One procedure which has been introduced recently for the relief of spasticity in cerebral palsy is that of selective posterior rhizotomy (Peacock and Arens 1982). In this operation the posterior roots from L2 to S1 are exposed and stimulated electrically:

> The responses in the muscles due to dorsal root stimulation fall into two groups. The type A or normal response is characterized by: (i) a single muscular contraction at 50 stimuli per second; and (ii) no diffusion of muscular contraction to muscle groups other than the one being stimulated. In the type B or abnormal response there is: (i) a tetanic muscular contraction at 50 stimuli per second; and (ii) a diffusion of muscular contraction to muscle groups other than to those being stimulated.

The roots showing type B responses are split into their rootlets which are then stimulated individually. Those showing A activity are left intact as they are not thought to be responsible for the spasticity, and so by saving them the afferent fibres useful for sensation and further motor re-education are preserved.

This major intervention is undertaken on the assumption that the hypertonia is due to stretch reflex activity. Our findings, however, show that this assumption is false, at least where calf musculature is concerned. Serial casting, or time-honoured treatment of tenotomy (where the Achilles tendon is lengthened) may well prove to be the most effective therapy. There is clearly room for more than one view, as reflected in the lively correspondence in the *Journal of Child Neurology*[10]. And as Brown and Minns (1989) emphasised, there are serious clinical problems:

> One should perform tendon lengthening only on a fixed equinus with a true contracture, otherwise excessive elongation of the tendo Achilles has secondary effects on the knee and hip as

[10]*Journal of Child Neurology* (1991) **6**, 173–180.

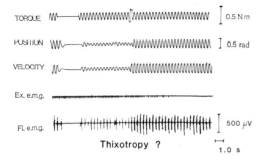

Fig. 10.33. Effect of a perturbation at the wrist in a hemiplegic patient. There is an increase in the motion after the disturbance, and associated EMG discharges in the forearm flexors.

Fig. 10.34. Rabies (hydrophobia). There are paroxysms with a sense of suffocation attended with a sudden convulsive heaving of the chest, catching of the muscles of breathing, and an inexpressible degree of agony in the countenance. In the engraving the patient views with horror the sight of water (from Bell 1824).

well as producing calcaneous deformity . . . Equinus deformity may be due to a spastic calf caused by a dynamic deformity from rigidity (the dystonic phase of cerebral palsy), it may be secondary to hip and knee flexion contractures, or it may be positional as a result of lying with tight bed clothes or when the child has been nursed on his face with hips abducted and the feet plantar flexed, as in the child with congenital tight heel cords. Equinus may also be due to a forefoot equinus which results from constant toe-walking.

Thixotropy
Because of the change of innervation in spasticity muscle properties change. In the extensor digitorum brevis, several months after a stroke, the number of motor units was reduced by about half while the remainder became larger and slower (McComas *et al.* 1973). It is likely that thixotropic effects will also be different. Thixotropic changes in disease have not been studied extensively. With spasticity, however, although the motion needed to elicit the responses is quite small, there may be EMG discharges at each swing. The contribution of passive properties is mixed with reflex effects (Fig. 10.33).

Spasms

Rabies
Spasms occur in a wide variety of conditions. The worst, which is still nonexistent in Britain, is rabies. Any afferent stimulant, a sound, a draught or the mere association of a verbal suggestion may cause a violent reflex spasm. The spasms

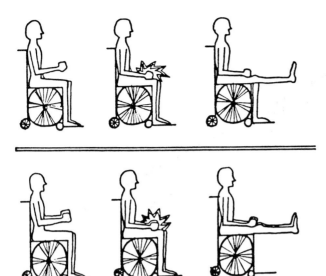

Fig. 10.35. Paraplegic patient. By striking his thighs he could induce a spasm of the knee extensors and so lift himself up into a standing position (from Harris and Walsh 1972).

which affect the muscles of the larnyx and mouth are exceedingly painful and associated with an intense feeling of breathlessness. Any attempt to take water is followed by an intensely painful spasm of the muscles of the larynx and elevators of the hyoid bone. It is this which makes the patient dread the very sight of water and gives the name hydrophobia to the disease (Fig. 10.34).

Paraplegia

A 28-year-old paraplegic, employed in a telephone control centre, used a trick to raise himself from the seated position. By thumping both thighs with his fists, an extensor spasm of the knees allowed him to stand with the aid of sticks or a little help from someone else. The mechanical blow appeared to set up a clonic contraction which merged into a continuous extensor thrust. The procedure only worked if he was lying back in his chair some 20° from the vertical. This may indicate that the rectus femoris, which runs over both hip and knee, was important in the reaction. If he hit only one thigh, only that leg extended (Fig. 10.35). When the knee was tested (as described on pp. 107–110), electrical activity was found in the quadriceps both during flexion and extension; the doubling of activity in the thigh acted as a 'full-wave rectifier', to use an electrical analogy. Thus movement either way could produce a contraction and the extensor state could be established.

Torticollis and head control

Observations were made on a limited number of patients with torticollis. Each subject was fitted with an alloy helmet and forces were transmitted from a printed motor through a linkage to stretch the sternomastoid muscle. No evidence of stretch reflex activity was seen on the EMG recording.

By repositioning the attachment of the linkage and the orientation of the

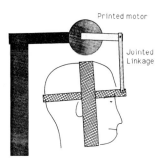

Fig. 10.36. Forces provided by a printed motor move the head in the sagittal plane.

motor it was possible to test the head's response to forces acting in other planes (Fig. 10.36). No resonance of the head was found, the motion merely becoming smaller as the frequency of the torque was raised. This stability is perhaps due to influence of the vestibular system reflexly on the neck musculature.

These observations were made with the head in its normal erect posture. For understanding the biomechanics it is desirable to know something about the centre of gravity:

> The head would be in a state of perfect equilibrium on the spine, in the erect attitude of our body, if the parts in front of the column exactly counteracted those behind it. This, however, is not the case. The articular condyles are manifestly nearer to the occipital tuberosity than to the most prominent point of the jaws; and thus the greater share of the weight is in front of the joint . . . The inclination of the head forwards is counteracted in the living body by the extensor muscles, and their constant exertion is necessary for maintaining the head in equilibrio on the vertebral column. Whenever their contraction is suddenly suspended, as in a person falling asleep in the erect attitude with the head unsupported, that part, abandoned to the force of gravity, immediately nods forwards. (Lawrence 1866)

This view is no doubt correct although the imbalance may be less than at first appears since so much of the front of the head is provided with air-filled cavities. The editor has added a discordant footnote:

> The weight of the head in sleep has a tendency to *preponderate equally in every direction*, as we see in those who are dozing in a carriage. Nay, their heads sometimes revolve in a circle, like the head of a harlequin on the stage.

Cramp

Mosso (1904) wrote about cramp:

> Some people find writing a few lines sufficient to exhaust their hand; they are forced to desist not only because the writing becomes illegible, but also on account of the pain, the tingling, and the sensation of tension which they experience in the muscles of the hand. When muscular cramp occurs in pianists and violinists, it forces them also to rest . . . There are very good swimmers who dare not go far from the shore, because they are afraid of cramp in their calves. We have all experienced the pain caused by cramp when it occurs unexpectedly at night during sleep. Usually it occurs after muscular contraction, but in very nervous people it may happen when the legs are immobile.

There are numerous types of occupational cramps. In these 'craft palsies', people who have been skilled and have had no problem for many years may find

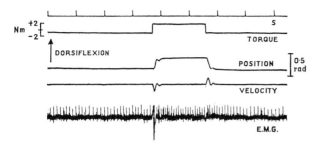

Fig. 10.37. Racing cyclist. Discharge of a motor unit in gastrocnemius after exercise which accelerates when the calf is stretched but also continues when the muscle is unloaded (from Muir *et al.* 1970).

that painful tightness occurs in a muscle group which is vital for their occupation: playing a sport or a musical instrument, cutting hair or painting.

During writing, too, many muscles are involved. When a small boy struggles to acquire the skill, his leg may be curled round the chair to steady him. The patient with writer's cramp in a sense regresses along the path which s/he once trod to learn the skill. The handwriting grows clumsier and the hand aches; and as the skill to use the appropriate muscle wanes, inappropriate muscles are invoked, normal postures of the finger or hand are adopted and obvious muscular spasm may occur.

The origin of these disabilities is obscure. It may be the first sign of idiopathic torsion dystonia. Treatment is unsatisfactory, and the person may have to give up the occupation involved.

Racing cyclists suffer from various neuromuscular aberrations, such as clonus, tremors, cramp and sudden loss of power.

A Scottish road-race champion, aged 24, had been training hard on the banked track of the Edinburgh velodrome just two minutes before his right ankle was tested (see Fig. 10.22). The EMG showed large single spikes in the gastrocnemius which persisted even though he tried to relax. When the muscle was stretched, using low-frequency abruptly alternating torques, the discharges were seen to accelerate; but when the motor plantarflexed the foot, the unit continued to discharge albeit at a slower rate (Fig. 10.37). This type of activity was never seen in normal subjects who had not exercised. The abnormalities persisted for two minutes after the recording was started and did not recur.

Another racing cyclist, aged 32, had just completed 25 miles (40km) training on roads on a cold wet day. Sinusoidal forces at 8Hz were being used during the procedure for testing the ankle and he was asked to try to resist the motion. When he tried the EMG, channels showed activity in antagonistic muscles and the foot was partially plantarflexed (Fig. 10.38). The oscillations resulting from the applied force were not reduced in size, presumably because the frequency of the torque was relatively high. After some 12s he was instructed to relax but found it impossible. The muscles continued in contraction and he complained of cramp. This persisted when the force was withdrawn. The muscles were kneaded manually and the cramp subsided after about a minute. The contraction had induced a state that was

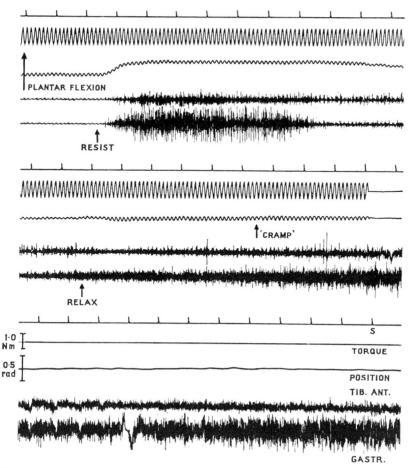

Fig. 10.38. Racing cyclist. Records are to be read as a continuous sequence from left to right. Initially rhythmic force applied to foot causes small movement. When he tries to stop the motion there is intense EMG activity and after some seconds cramp develops. This is neurogenic rather than myogenic since the presence of large spikes indicates the activity of motor units rather than individual muscle fibres (from Muir *et al.* 1970).

difficult to reverse. In both these cyclists, involuntary contractions occurred which could not be stopped by attempts at relaxation.

Cramps are not always associated with electrical activity. In McArdle's disease a major symptom is limitation of vigorous activity because of cramps after exercise, during which the muscles are electrically silent. There is pain and stiffness of any muscle on moderate exercise and there may be marked swelling. After powerful gripping movements the fingers may remain partially clenched for several minutes. Climbing stairs or rapid walking may lead to excessive breathlessness and an inordinately rapid pulse, both probably due to an excessive increase in muscle bloodflow.

In one person affected with the disease, a medical student, a small piece of living muscle was excised for study. Single fibres stripped of their sarcolemma behaved abnormally in that after several challenges by caffeine or calcium they went into contracture for up to a minute. In normal muscle the same chemical stimuli gave effects lasting only about 2s (Gruener et al. 1968).

The muscle membrane is believed to be normal: the problem lies in the 'excitation-contraction coupling'. The condition is metabolic; because of a lack of muscle phosphorylase there is an inability to use muscle glycogen.

Hypertonia during and after anaesthesia
Ketamine is an unusual agent which gives a form of 'dissociated anaesthesia'. There is complete analgesia with only superficial sleep. The patient may gaze sightlessly into space without blinking for several minutes. The eyelash and corneal reflexes remain unimpaired and there is usually increased muscle tone accompanied by grimacing or involuntary movements, but no response to auditory stimuli.

It has long been known that morphine may induce 'catalepsy' in rats. More recently knowledge of the production of rigidity by opiates has become relevant to clinical practice. This is because large doses of narcotics are sometimes used as anaesthetic agents. They are used particularly in high-risk patients because they generate only a minimal load on cardiopulmonary function.

The rigidity produced by one of the compounds, alfentanil, was studied by Benthuysen et al. (1986):

> Rigidity and EMG activity were often provoked or increased by stimulation of the patient, for example, by passive movement of an extremity, manipulation of the mask, or loud auditory stimulus. Clinically, rigidity was often explosive in onset, with the subjects assuming postures typified as follows: flexion of the upper extremity at the fingers, wrists, and elbows; extension at the toes, ankles, knees, and hips; rigid immobility of the head with atlanto-occipital flexion of the chin onto the chest; and severe rigidity of the abdominal and chest wall musculature. In two patients receiving alfentanil, ineffective inspiratory efforts were observed after 2–3 min of apnea when rigidity was still present. Following apnea . . . extension of the neck and insertion of an oral airway were impossible. At this time, two individuals were required to ventilate the patient; one to maintain a mask fit and another to apply positive pressure. Relaxants completely eliminated all clinical as well as EMG evidence of rigidity, and no difficulties in ventilation or intubation were subsequently encountered.

The authors observed that rigidity still developed in the arm when the circulation was occluded by a tourniquet, so it is clearly central, not muscular in origin.

After any generalised anaesthetic, but particularly after halothane, there is usually a period when muscle tone is high. This was studied by Soliman and Gillies (1972):

> Spasticity occurred when the patient had started to respond to painful stimuli and disappeared when the patient was responding to verbal stimuli. Its duration averaged 6–7 minutes and depended on the rate at which the patient recovered consciousness. Ankle clonus could be elicited during the period of spasticity and was most sustained when spasticity was most intense. Extensor plantar responses were present during the early stage of spasticity.

Rare, and potentially disastrous, is the development of malignant hyperpyrexia in association with anaesthesia. There is a rapid rise in body temperature and muscle stiffness. In the very small number of people with this inherited condition, halothane and suxamethonium are very dangerous. The treatment is by the drug dantrolene, which reduces calcium release in the muscle fibres. Dantrolene has been used as a muscle relaxant for other conditions, but it is questionable whether it can be recommended.

Shivering
Shivering depends on the integrity of the hypothalamus but not of the forebrain. It is mediated by the 'extrapyramidal' system as it is shown that following a stroke, on cooling, shivering may invade the paralysed limbs. A limited number of observations were made of wrist resonances in volunteers being cooled. One of the subjects, being treated with reserpine for hypertension, coincidentally became prone to shiver when encountering trivial degrees of cooling. There was no increase in resonant frequency before the shivering commenced (*i.e.* no pre-shivering increase of tone). The metabolic rate may increase in cold conditions without change in muscle tone, because some animals, infants and perhaps certain adults have 'brown fat' which increase its metabolism in response to circulating catecholamines.

Shivering develops independently in each muscle group. The character of the movement changes according to time and place in the limbs. When the cooling is intense, however, the rhythm becomes identical in almost the whole musculature. This shuddering implies the presence of a coordinating mechanism to secure the similarity in widely spaced parts.

Patients who have been given a general anaesthetic frequently develop tremulous movements of the musculature as they are recovering, just after the period of hypertonicity discussed above. The movements, which can follow the use of a wide variety of different anaesthetic agents, are usually regarded as a form of shivering since some drop of body temperature commonly occurs. It is highly undesirable: there may be a large increase of metabolic rate, disruption of surgical repairs and damage to teeth. Using the lung as a heat exchanger is useful for maintaining body temperature. Pflug *et al.* (1978) pointed out that:

> Pulmonary heat transfer can take place over a large surface area which represents 50 to 70 square meters of alveolar epithelial-pulmonary capillary interface in the adult and a mucosa to bronchial blood interface in about one million small bronchioles. This air-tissue and tissue-blood heat transfer takes place at the center of the body in an organ that surrounds the heart and the warmed blood then moves to the peripheral tissues through the vascular system.

It was found that 50 per cent of unwarmed patients shivered postoperatively; none shivered who had been warmed.

Grasp reflexes, catatonia and catalepsy
The normal baby shows a grasp reflex. If an object of suitable size is placed in contact with the palm, the fingers close and there is resistance if an attempt is made to pull it out. If grasp reflexes are elicited on both sides the baby may be lifted. As

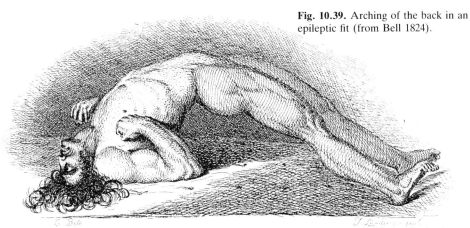

Fig. 10.39. Arching of the back in an epileptic fit (from Bell 1824).

the infant matures the grasp reflex is lost but it may reappear with brain lesions.

In a form of schizophrenia, catatonia, unusual postures may be held for long periods. This condition is quite rare in Britain but is said to be common in India. The author is not aware of any neurophysiological investigations of this condition.

An alkaloid, bulbocapnine, has been used to induce changes which some have interpreted as related to a condition in animals, catalepsy. A posture may be adopted by a cat treated with this drug and the animal can remain essentially motionless in that position for a long time. The clinging position which may be shown depends on proprioceptive information from the forelimbs, the animal holding on with its forepaws to a support such as the back of a chair (Harreveld and Bogen 1961).

Stiff babies, stiff men and episodic hypertonia
In babies there is an autosomal dominant condition in which there is an onset at birth of generalised hypertonia in flexion, which disappears in sleep. The children show exaggerated startle reactions to trivial stimuli. There is acute diffuse hypertonia which makes them fall like a log, mostly onto the face, without clouding of consciousness or epileptic phenomena. This 'stiff baby' syndrome is known as hyperexplexia. There is delay in reaching normal motor milestones such as walking. The hypertonia is reduced during the first two years but involuntary starts to the slightest stimuli persist (Kok and Bruyn 1962).

The 'stiff man' syndrome develops after the age of 40. Evolution is slow and progressive, the muscles show contractures and there is dysphagia. Here, too, trivial stimuli increase the hypertonia. There is no family history.

Muscle contractions of a tonic then clonic type occur in the *grand mal* form of epilepsy (Fig. 10.39).

Apart from poisoning, as by strychnine, there are other conditions where muscle tone may increase episodically. There are the 'Jumping Frenchmen of Maine'. This condition usually has an onset in adolescence. These people are shy and abnormally ticklish. They show exaggerated startle reactions in which they may jump, scream, throw an object they are holding, hit, swear and obey shouted commands. Saint-Hilaire *et al.* (1986) recounted that:

> An old lumberjack recalls that on the first day of work somebody would yell to discover who was the jumper in the group. This was not a rare condition, and there was at least one jumper on almost every site. The only woman we studied had many jumpers in her family who were also working as lumberjacks, and they often startled each other for amusement.

Similar groups have been reported from other countries.

Nerve injuries

Naturally with nerve injuries there are disturbances of muscle function. Weir Mitchell (1872) made careful investigations into the effects of gunshot injuries at the time of the American Civil War:

> I have already described the sudden and violent form of spasm which is sometimes the first symptom of the wound of a nerve by a ball, and I have also called attention to the painful spasms which may result almost instantly from the wounding, and probably even from the division, of a small sensitive skin nerve in bleeding at the elbow. The latter variety of spasm may last for days, and even for weeks, until, as in one instance, the growing nails of the clinched hand may cause ulcers in the wounded palm. In these cases, as in wounds from missiles of war, the spasms are of reflex origin, and are often seen in the muscles of unwounded nerves. In the case of J. H. C. the bullet injured the left median and ulnar nerves, and caused instant cramp of the flexors of both arms, so that he clutched both gun and ramrod violently. He shook loose the latter with a strong effort, and then, with the right hand thus set free, he unlocked the fingers of the left from their clutch on the gun, after which they did not again close.

Tetany

A maintained contraction of muscles occurs in this condition, caused by either a reduction of serum-ionised calcium concentration or by over-breathing. This name and that of tetanus comes from the Greek τείνειν, 'to stretch'. The EMG consists of single or often closely spaced pairs of spikes which occur with machine-like regularity (Kugelberg 1946). The rigidity is plastic in that if the hand is bandaged into a particular position and then tetany induced by hyperventilation it is this position which is retained when the bandage is removed (Rosett 1924). Without bandaging, the hand takes up the position known as '*main d'accoucheur*', because it is similar to the posture that may be useful in practical midwifery.

Tetanus

Tetanus, particularly of the newborn infant, is relatively common in the Third World because villagers live close to the soil and immunisation is frequently not undertaken. Observations were made on patients in India at the tetanus ward of the Baroda Hospital, Gujarat. There develops in this condition intensive muscular hypertonicity. The colloquial name, 'lock-jaw', refers to the problem when the person attempts to open the mouth but the jaw closing muscles come powerfully into play preventing the action. EMG recordings in adult patients confirm that there is a failure of reciprocal inhibition in this condition (Fig. 10.40).

A phrase used by physicians to describe the macabre grin of patients with

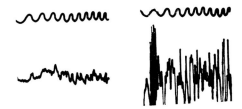

Fig. 10.40. Young adult Indian woman with tetanus. *Right:* lying with her head on the pillow there is slight action in the trapezius muscle. *Left:* on trying to raise her head the trapezius becomes active and powerfully resists the movement she is trying to make, a failure of normal reciprocal inhibition. Oscilloscope time marker 10Hz.

tetanus is *'risus sardonicus'*. The expression has been used since the time of Homer. *Risus* is Latin for 'laughter', but the etymology of the other half is debatable: it may refer to a poisonous hairy buttercup (*Ranunculus Sardous*, found in Sardinia amongst other Mediterranean countries), the eating of which causes the person to pull a wry face, or to the grin of a dog[11]. The root *sard* may refer to a place in Asia Minor[12] and the belief that burns suffered by smelters of metal in the Bronze Age in that region allowed the bacterial spores ready access to the body. The origin may perhaps be associated with Talos, the god of the smiths (Rosner 1956).

Myositis ossificans

In myositis ossificans, heterotopic new bone is formed and part of a muscle may calcify. This may follow trauma, sometimes of an apparently trivial nature, but often there is no history of damage or illness. It is commonest in young people. Physiotherapy should be undertaken very cautiously as stretching may aggravate the condition. Myositis ossificans is quite unusual, however, and for most people the ultimate hypertonia is rigor mortis!

Rigor

In a study of mechanisms concerning the filaments of muscle fibres, Huxley and Brown (1967) gave his account:

> Muscles in rigor (when the cross-bridges are permanently attached to the thin filaments) give X-ray diagrams which differ very considerably from those from resting muscle. The changes take place largely in the myosin component and the results indicate that a co-operative re-organization of the helical arrangement of myosin cross-bridges may occur when they bind to the sites on the actin filaments in such a way as to maximize the number of points of near-registration. The actin filaments appear to behave as relatively invariant structures though small changes in pitch cannot be excluded.

Individual muscle fibres stiffen in a short time, about two minutes, and the concentration of calcium ions greatly increases. Differences between different fibres, however, mean that the stiffening process in a whole muscle may be spread over quite a long period. There is great variation in the development of stiffness after death. The causes were investigated in rabbits by Bate-Smith and Bendall (1949), who devised an instrument for measuring the modulus of elasticity of psoas

---

[11]Perhaps from σαίρω (sairo)—'the parting of the lips in a grin'.
[12]There are Biblical references to Sardis in Revelations I.11 and III.1–6.

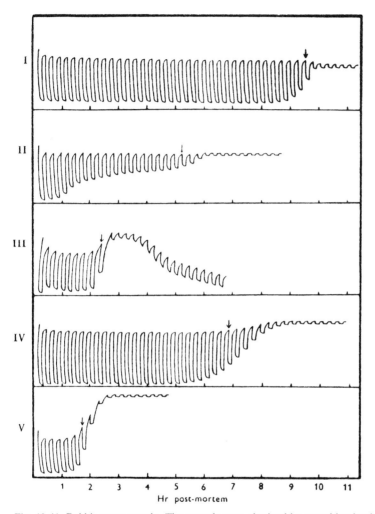

Fig. 10.41. Rabbit psoas muscle. The records were obtained by stretching it with a 100g weight, 8 min on, 8 min off. With the development of rigor mortis the stiffness increased 20 fold. The time course is different in each of the five examples (from Bate-Smith and Bendall 1949).

muscles. This muscle has two advantages for the measurement of stiffness: the fibres are relatively straight, and the muscle is exceptionally free of connective tissue. Examples of the records are shown in Figure 10.41. Two factors were identified. If the animal had been well fed and struggles at death had been prevented by the use of a neuromuscular blocking drug, onset was delayed. If the animal was glycogen-depleted, however, initially the onset was more rapid. Low glycogen reserves and a struggle led to a rapid stiffening. The factors were related to the pH of the muscle; as rigor sets in the muscle becomes acidic.

Meat is treated by refrigeration; carcasses are cooled rapidly after slaughter.

This is necessary to prevent the growth of bacteria but can give problems as pointed out by Winstanley (1986):

> Unfortunately, rapid chilling of carcasses before the meat has entered rigor produces extremely tough meat—a phenomenon known in the meat trade as 'cold shortening'. As its name implies, cold shortening involves muscle contraction. Muscles from a freshly slaughtered carcass can shorten to less than half their resting length if they are cooled rapidly to below 10°C. Even when muscles remain fixed to the skeleton at both ends you can see zones of marked cold shortening, with stretching elsewhere.

Storing meat makes it more tender. This is called 'conditioning' or more familiarly 'hanging the carcass'. The change involves rupture of the z lines, the transverse structures which separate adjacent sarcomeres. Lysozymes—enzymes normally in the tiny membrane sacs—are released after death and connective tissue is broken down.

For medico-legal purposes, the time of onset or rigor may be important. An account of the development in man was given by Howell (1905):

> After the death of an individual the muscles enter into rigor mortis at different times. Usually there is a certain sequence, the order given being the jaws, neck, trunk, upper limbs, lower limbs, the rigor taking, therefore, a descending course. The actual time of the appearance of the rigidity varies greatly, however; it may come on within a few minutes or a number of hours may elapse before it can be detected, the chief determining factor in this respect being the condition of the muscle itself. Death after great muscular exertion, as in the case of hunted animals or soldiers killed in battle, is usually followed quickly by muscle rigor; indeed, in extreme cases it may develop almost immediately. Death after wasting diseases is also followed by an early rigor, which in this case is of a more feeble character and shorter duration.

'Cadaveric spasm' has an almost immediate onset in one limb after violent death. It is beloved of crime writers but very rare. The person may clutch an object in a vicelike grip. If he has committed suicide, the gun or knife may be grasped. If there has been a murder, the victim may retain in the hand a button or piece of hair of the assailant.

11
WHOLE BODY VIBRATIONS AND INTEGRATIVE ASPECTS

In the preceding chapters, much space has been devoted to resonances in different parts of the body. It is now appropriate to consider some of the possible interactions, how these may be affected by variations of muscle tone, and relationships to voluntary control.

Force transmission: the tendons

The role of muscles has been considered, but it is also relevant to review the functional significance of tendon. The mechanical properties have been reviewed by Evans and Barbenel (1975). The tensile strength is in the range of 5 to 10kg/mm^2. Tendon is metabolically inert, slow to repair, slow to degenerate and 'virtually dead, even during life'. It is a flexible force-transmitting element. John Hunter (1837) wrote:

> It is composed of white fibres placed parallel to each other, forming a chord, which is extremely flexible, has no sensible elasticity, and is much smaller than the power to which it is attached. Its figure is in general a little rounded; sometimes, however, rather flattened, and in many situations it is broad and thin; in all cases it is extended between the body to be moved and the power . . . It intervenes between the body to be moved and the power to keep up the exact proportion necessary between them to produce any determined motion, so that the length of the bones, or the distance between the joints or points of motion, the quantity of motion in the joint, and the quantity of contraction in the muscle, are proportioned to one another. But if this substance had been wanting, and the muscular fibres had extended the whole distance between many of those joints, the power of contraction in such a muscle would have been too great, especially in the extremities . . . A most beautiful and forcible example of the use of tendon in limiting the length of a muscle to the extent of motion required to be produced in the part to be moved occurs in the sterno-thyroidei of the giraffe. Had these muscles been continued fleshy as usual, from their origin, through the whole length of the neck, to their insertion, it is obvious that a great proportion of the muscular fibres would have been useless, because such a condition of the muscle would have been equal to have drawn down the larynx and os hyoides more than one third of the extent of the neck, which is neither required nor permitted by the mechanical attachments of the parts . . .
>
> It was necessary that some substance should be introduced as a medium between bones and muscles, to admit of the nicety of action and freedom of motion we find in many parts of the body, particularly in the fingers; which could not have taken place if the muscles had been continued from bone to bone.

In speedy animals (*e.g.* the horse and camel) it is found that the main muscles moving the limbs are in or close to the trunk; the peripheral musculature is largely vestigial. The forces are transmitted by long tendons. By this arrangement the moments of inertia of the legs are comparatively low and rapid strides can thus be made. The fingers of the mole are flexed by the pectoralis and teres major, while the bird's wrist is extended by the triceps.

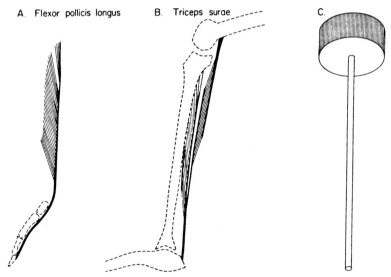

Fig. 11.1. *A:* The long tendon of the human flexor pollicis longus. *B:* The arrangements for the human gastrocnemius and soleus. *C:* The muscle fibres and tendon of the gastrocnemius and soleus have been rearranged to show their relative lengths and cross-sectional areas. (From Rack 1985, by permission.)

Tendons are often looked upon as being inextensible but Rack (1985) pointed out that:

> In larger animals (such as man), muscle fibres are often considerably shorter than their tendons of attachment, and the sum of the muscle fibres has an effective cross sectional area that is much larger than that of the tendon. These anatomical features are obvious in some of the forearm muscles whose tendons extend down into the digits. Many other limb muscles also have surprisingly long tendinous attachments though they do not always appear in well-defined tendons at one or other end of the muscle.

Rack supported his arguments diagrammatically (Fig. 11.1). In different animals there are natural variations in the proportion of a muscle given over to tendon. Alexander (1989) observed:

> The most extreme adaptation is seen in the plantaris muscle of the camel: this muscle runs from the back of the knee to the toes and measures 125 centimetres in an average one-humped camel. This long 'muscle' has lost its muscle fibres, apart from some useless rudiments about two millimetres long. But a tendon as thick as a finger still runs from knee to toes. So during the course of evolution the muscle has been replaced by a tendon that serves as a passive spring. In this exceptional case, the best muscle is no muscle at all.

During running, if active muscles are being stretched, energy is stored in the tendons and can be released later. The kangaroo hops along like a bouncing ball. In this gait energy is stored in the tendons on landing and then released as the mass of the body moves upwards again (Alexander 1988).

Gear ratios

Some muscles are attached close to the axis of the joints about which they act,

Fig. 11.2. 'Ocydromes', swift runners from a Greek vase (from Marey 1895).

Fig. 11.3. Nineteenth-century method of recording movements of a man walking and running (from Marey 1874).

others are inserted further out. 'Cursorial' animals, which run rapidly, have special adaptations. Hildebrand (1960) commented:

> Cursorial animals not only have longer legs; their actuating muscles are also attached to the bone closer to the pivot of motion. Their high-gear muscles, in other words, have short lever-arms, and this increases the gear ratio still further. In comparison, the anatomy of walking animals gives them considerably lower gear-ratios; digging and swimming animals have still lower gear ratios. But while high-gears enable an automobile to reach higher speed, they do so at the expense of power. The cursorial animal pays a similar price, but the exchange is a good one for several reasons. Running animals do not need great power: air does not offer much resistance even when they are moving at top speed. Moreover . . . the animal retains some relatively low-gear muscles. Probably the runner uses its low-gear muscles for slow motions, and then shifts to its high-gear muscles to increase speed.

Horses attain high speeds also by moving a number of joints in the same direction at the same time. Instead of having the foot on the ground the horse is standing on hypertrophic toes, in effect, and extra limb joints have thus been obtained. Still more movement is achieved because the shoulder blade can swivel. Large cats such as the cheetah, too, increase their speed by alternately flexing and extending the back.

Whole body vibrations of low frequency

The study of vibration looms large in mechanical engineering; problems as diverse as the design of bridges, the balancing of petrol engines and the effects of oscillations on aircraft structures are obvious examples. But what do jolts and vibrations do to human beings? Do mechanical oscillations cause physical and mental fatigue?

The short answer is that the effects vary according to the vigour of the thrusts, doing no permanent injury but capable of producing considerable discomfort. Mild vibration—produced by riding a horse or in a 'bone-shaker' motor car—can be exhilarating. Walking or running subjects the body to rhythmic forces of considerable power and the result may be a sense of well-being.

Oscillations induced by human locomotion

Substantial rotary forces are involved in human locomotion (Figs 11.2, 11.3). In walking, the arms swing at the same frequency as the strides. This swinging is not merely the passive action of pendula but is power-driven to a significant degree by muscular action. According to Elftman (1939):

> The swinging of the arms serves to regulate the rotation of the body as a whole, serving especially to decrease rotation about the vertical axis while only one foot is on the ground and to modify the rotation while both feet are in contact. The rotation of the body about the direction of progression is modified by the arms so that the change in direction of the rotation is accomplished more smoothly. The magnitude of these effects depends on the amplitude of the swing, since this affects the velocity, and of the positions through which the arms swing, since they affect the distance of the parts from the center of gravity of the body.

Swinging of the arms was also considered by Roberts (1967):

> In man, the moment of inertia about the central vertical axis is much less than that about any other axis, because of the erect posture. Consequently, the unsymmetrical forces involved in locomotion can give rise to considerable angular accelerations about the vertical axis. By swinging the arms and shoulders backwards and forwards in contrary motion to that of the legs, the trunk is set into a torsional oscillation from which angular momentum is available at the appropriate phases to counteract the effect of the rotational forces generated by the legs. In this way, it is possible to keep the head pointing in roughly the same direction during the progression, in spite of the fluctuating torsional forces applied to the trunk by the legs.

Pettigrew (1873) stated:

> It will be evident that the trunk and limbs have pendulum movements which are natural and peculiar to them, the extent of which depends upon the length of the parts. A tall man and a short man can consequently never walk in step if both walk naturally and according to inclination. In traversing a given distance in a given time, a tall man will take fewer steps than a short man, in

the same way that a large wheel will make fewer revolutions in travelling over a given space than a smaller one . . . When a tall and short man walk together, if they keep step, and traverse the same distance in the same time, either the tall man must shorten and slow his steps, or the short man must lengthen and quicken his. The slouching walk of the shepherd is more natural than that of the trained soldier. It can be kept up longer, and admits of greater speed. In the natural walk, as seen in rustics, the complementary movements are all evoked. In the artificial walk of the trained army man, the complementary movements are to a great extent suppressed.

Injuries caused by shaking infants

The most common reason for repeated whiplash-shaking of infants and young children is to correct minor misbehaviour, but 'a good shaking' can result in serious damage to brain, eyes, and limbs, with subdural haematomata and mental retardation. X-rays of children abused in this way have shown small fragments of bone torn away and small chunks of calcified cartilage adrift from the growing ends. This avulsion appears to result from the acceleration–deceleration forces rather than direct impact stresses on the bone itself (Caffey 1972).

In the Royal Hospital for Sick Children in Edinburgh, non-accidental injury was the third most common single cause of acquired hemiplegia in the early 1960s. Shaking injuries are an important cause of severe morbidity and mortality in babies.

Head control, woodpeckers and concussion

African women often carry prodigious loads on their heads in a manner which is metabolically very efficient. The metabolic work involved was studied by Maloiy *et al.* (1986) who asked:

> How are the African women able to carry such large loads on their heads so economically? It seems reasonable to assume that the energy requirements for moving the legs and arms relative to the centre of mass would be the same for loaded and unloaded people because stride frequency is unchanged when carrying a load. However, the energy requirements for lifting the centre of mass within each step, and for accelerating and decelerating the centre of mass at each step, would be proportional to the total mass (M_b[1]+load) if the movements of the centre of mass were unchanged; the African women may be more economical simply because they are able to minimize the movements of the load when supporting the load with their heads (the available experimental evidence shows that the amplitude of vertical movements during normal unloaded walking decreases as one moves up the spine from the coccyx to the head). The fact that the African women alone can carry loads of up to 20% M_b without any measureable metabolic cost may indicate that some anatomical change has occurred, as a result of carrying large loads since childhood, which allows these women to support small loads using non-metabolizing structural elements.

One of the striking aspects of posture is the way the head is stabilised. This is particularly clear in birds. The head is motionless between movements, and a buzzard quartering the ground below for prey will hover with its head quite still. When a pigeon walks, its head does not really bob backwards; it just appears to do so. The head remains still between forward jerks.

Our limbs swing as we walk but the head is steady; it acts as a stabilised

---

[1]Body mass.

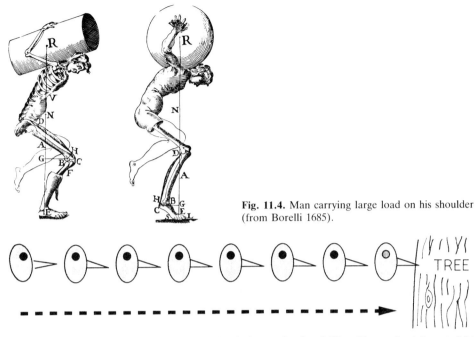

Fig. 11.4. Man carrying large load on his shoulder (from Borelli 1685).

Fig. 11.5. Movements from left to right of the head of a woodpecker drilling. The motion is in a straight line and must be a complex accomplishment as a number of different joints must be involved. The eyelids close about 2ms before impact, they will both protect the corneas from chips and act as 'seat belts' preventing the eyes popping out. Based on cine-film frames, but freely adapted, from May (1979).

platform from which the eyes can operate. Carrying a load on the shoulder is less sensible (Fig. 11.4).

Some ornithologists have asked why woodpeckers do not knock themselves out when they drill. On each forward movement—about 20/s—high velocities are reached (*e.g.* 7m/s, 27km/h, or 16mph). Such high speeds at such a frequency could only be achieved with resonance: energy in one swing is carried over towards the next. The rebound must be considerable.

The problem was investigated by May *et al.* (1979) using high-speed cinematography. A boxer, and apparently also a gannet diving into the sea to catch a fish, holds the neck rigid at the moment of impact. Goats have powerful neck muscles and these evidently operate to reduce rotation when they butt each other with their horns. Neck stiffening apparently occurs with the woodpecker, reducing the possibility of injury due to rotation. The drilling is achieved by the head moving in a remarkably straight trajectory (Fig. 11.5). The authors pointed out that:

> A linear trajectory, however, necessitates a much finer and *graded* adjustment of the whole postural apparatus, carried out continuously at high speed and involving a much more elaborate integration of both exteroceptive and interoceptive information.

The mechanics of head injuries were investigated by Holbourn (1943). There

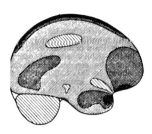 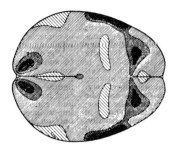 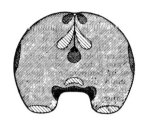

Fig. 11.6. Shearing by rotory acceleration of a model brain made of gelatin. The darker the area the greater the strain. *Left:* forwards rotation resulting from a blow on the occiput. There is damage at the tip of the temporal lobe as the skull gets a good grip on the brain in this region. *Middle:* effect of rotation in the horizontal plane caused by a blow on the upper jaw. *Right:* rotation in the coronal plane caused by a blow above the ear. (From Holbourn 1943, by permission.)

may be local injury due to skull distortion, and the other factor considered was the effect of skull rotation damage, which would occur even if the skull was absolutely undeformable. The brain is virtually incompressible and of almost uniform density but not rigid. It is not rise of pressure that is damaging but shearing due to acceleration of a rotatory type which is relevant (Fig. 11.6). Shear strain, or slide, is the type of deformation which occurs in a pack of cards when it is deformed from a neat rectangular pile into one obliquely angled. If the tissues are separated too far they suffer potentially irreversible damage.

Horses

Robert Hooke (1635–1703) described experiments on respiration, developed compound microscopes and noted the cellular structure of plants. He has been considered as the 'greatest of the philosophical mechanics'. He is remembered especially at being a student of the properties of springs. Hooke's law states that '*ut tensis sic vis*'—or 'as the stretching so the force'. In other words, if a spring is pulled out, the tension is proportional to the extension. Hooke experimented with a sprung saddle for horse riding, and such things are still manufactured. Horse-riding can only be enjoyable when the rhythmic movements of the horse and rider are matched.

Roberts (1985) considered the head movements of the horse:

> The head of the horse is relatively very heavy because, although it contains air spaces in the nasal cavities and elsewhere, it also contains the large masses of ivory forming the grinding teeth. The heavy weight at the end of the long neck forms a massive pendulum which needs large forces to set it in motion and to change its speed or direction of movement. The muscles that pull on the neck to move the head also pull on the trunk. In this way the horse can use the momentum of its very heavy head to assist in moving other parts of the body . . . For example when the horse lifts his head the muscular effort involved has the indirect effect of tending to lift the hindquarters, because the support at the shoulders acts as the fulcrum of a lever. When the horse is walking briskly you will see that he lifts his head each time he needs to bring one of his hindlimb legs forward.

Watching a horse moving in its various gears, and hearing the thud of the hoofs

Fig. 11.7. A trotting horse furnished with different experimental devices, the horseman carrying the instrument registering the pace. On the withers and the crop are sensors to show the reactions (from Marey 1895).

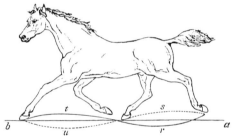

Fig. 11.8. Trotting horse. The legs describe a curved track and there is a torsional motion of the trunk. The originator of this drawing, an FRS, an eccentric genius and one-time curator of the Museum of The Royal College of Surgeons of Edinburgh, became famous for this description of a figure-of-eight configuration in the cardiac musculature and in the wing movements of birds and bats (from Pettigrew 1873).

on the ground as a horse thunders past in a canter or a gallop, one is struck by the rhythmic action which must surely be achieved by control of bodily resonances.

The gait of horses has interested many people[2] including the eminent French physiologist E.J. Marey (1830–1904). He used graphic methods of recording to study a wide range of problems including those of bird flight and of the cardiovascular system. Two of Marey's illustrations are reproduced (Fig. 11.7). Because the thrusts at the shoulders and hips are not synchronised and symmetrical the trunk executes a complicated motion. During a horse's movement there is twisting of the body and a spiral overlapping of the legs (Fig. 11.8). Roberts (1985) noted that:

> In the trot, the horse springs from one diagonal pair of legs to the other, making two bounces to a stride. In the canter there is only one bounce to each stride and the single unsupported phase correspondingly lasts longer than either of the two unsupported phases of the trot. A stronger thrust is needed to launch the weight of the horse's body.

[2] Most famous of all the investigators was the 19th century photographer Muybridge. By obtaining a series of pictures he was able to settle a dispute about the movements of the animal's legs. He used 12 separate cameras which could be exposed in sequence in half a second (Muybridge 1957).

When the man and trotting horse are not correctly 'phase-locked', the result can be distinctly uncomfortable. The natural period of the man's vertical movement is longer than the oscillations of the horse. The man usually comes down alternately on his seat and on his knees—so his total cycle time is double that of the horse. Rather exact timing is needed for comfort. Riding a camel travelling at a fast pace can be decidedly unpleasant: the fore and hind legs 'pace' *i.e.* they move together and there is a rocking motion.

Vibration induced by transport
Hooke was concerned with comfort in horse-drawn carriages and the role that springing could play in cushioning the shocks the passengers received when riding over the poor-quality roads of the times. The word 'travel' was originally the same as 'travail', meaning touble, hardship or suffering.

In Shakespeare's *Comedy of Errors* (I.ii.15), Anthipholus says:

> With long travel I am stiff and weary.

The springing of coaches was a new idea in the 17th century, and trials of sprung carriages are mentioned by Pepys. They remained uncomfortable; in the early 19th century the painter Cotman travelled in a rickety carriage in Normandy and 'it shook the very pockets off his coat'.

Piston-type military aircraft shook very badly. Coermann (1946), who worked for the German Air Force in World War II, studied the physiological effects and noted that:

> The right to be comfortable, in the view of many aircraft constructors, belongs only to the paying passenger. In the Luftwaffe, any attempt to provide a greater degree of comfort for aircrews is regarded as unmilitary.

Early railway carriages shook excessively; the carriages were cattle trucks and had no springing:

> Riding on an empty railway truck, loose coupled, with short springs, and having no buffers, conveys to the passenger a very accurate account of every inequality and irregularity of the line, or defect in the wheels. (*Anon 1862*)

The expression 'shaking the very life out of you' seems to have arisen in this context. With vibration there may be induced a state of muscular hyptonia through direct action on the muscle substance, through changes in thixotropy and through presynaptic inhibition as shown by H-reflex studies (pp. 111–114). It might be important to ascertain the relative importance of these very factors in assessing the most useful therapy for different categories of patients. It is related that Sir Henry Head, the eminent neurologist, found in his later years that the rigidity of his Parkinsonism was reduced when travelling in London in a hansom cab over cobbled streets.

Third-class passengers frequently suffered for days from the constant knocking together of the knees which successive jolts during the journey produced, for the

Fig. 11.9. 'The Third Class Carriage' by Daumier. Redrawn by F. Vance using the airbrush technique after an oil painting in the National Gallery of Canada, Ottawa, for whose cooperation we are grateful. A similar work is held by the Metropolitan Museum of Art, New York.

close packing prevented any wide separation of the legs (Fig. 11.9). The shaking could occur to a damaging extent. In the report of the commission on the influence of railway travelling on public health (*Lancet*, 8th February 1862), an 'eminent hospital surgeon' is quoted as follows:

> At a short period after the opening of the railroad from Leipsic to Berlin (I think it was in the summer of 1841), I was travelling in a first-class carriage to the latter city. I had for my companion a very corpulent man, upwards of sixty years of age, formerly an officer of rank in the Prussian army. The train was lightly laden and the carriages loosely coupled, and we had not proceeded far before we found the motion of the carriage most inconvenient, and, indeed to my fellow-traveller most distressing, in consequence of the shaking of his enormous abdomen. I placed him in the centre compartment of the carriage, persuading him to place his feet firmly against the opposite seat, packed him in his seat with great coats, &c., but in vain. His cries were piteous, and his aspect, as we approached the end of our journey, really alarming. For the last four or five hours I sat opposite to him, at his request, endeavouring to prevent his pendulous belly swaying from side to side with the motion of the carriage.

The same source (11th January) gives a good overall account of the effect of carriage oscillations:

> The tendency of each movement is to produce more or less motion in the twenty-four pieces of which the spine is made up. These movements are counteracted, and the erect position of the body is maintained, by the adaptive contraction of that complex muscular system attached to these osseous pieces. The more violent movements of the carriage call into action the various sets of muscles in the back and chest; and it is only by an incessantly varying play of contraction and relaxation that the body is preserved in a tolerable state of equilibrium, and that the passenger combats with success the tendency to be shaken into a most unpleasing variety of shapes and positions. The head is especially thus affected, being so balanced on the spine as to have a tendency to fall forward. The frequency, rapidity, and peculiar abruptness of the motion of the railway carriage keep thus a constant strain on the muscles; and to this must be ascribed a part of that sense of bodily fatigue, almost amounting to soreness, which is felt after a long journey.

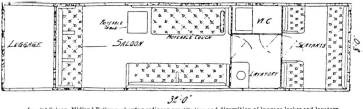

Invalid Saloon, Midland Railway, showing ordinary constitution and disposition of luggage locker and lavatory.

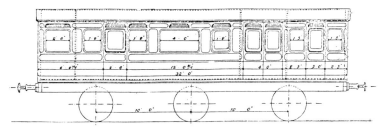

Diagram of the Invalid Railway Saloon, London and South-Western Railway.

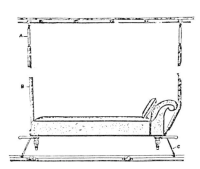

Invalid Saloon, South-Eastern and Chatham Railway, showing the method of slinging the invalid's couch.

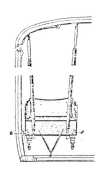

Invalid Saloon, South-Eastern and Chatham Railway, showing method of restraining oscillation of couch.

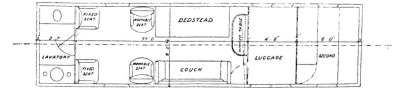

Invalid Saloon, London and South-Western Railway, showing alternative disposition of lavatory and luggage locker.

Fig. 11.10. Invalid railcars at the turn of the century. In the invalid saloon of the South-Eastern and Chatham railway the invalid's couch was suspended by cords from the ceiling. It is highly questionable as to whether such an arrangement would have been effective. Suspending a pendulum from a moving support does not, in general, lead to undisturbed motion; at times the movement may be exaggerated. An expensive mistake was made by Sir Henry Bessemer who had built a cross-channel steamer with swinging saloon. The assumption was that this would remain vertical whilst the ship rolled. It was a disaster (from Corner 1901).

Fig. 11.11. *Upper traces:* optical records of a railway passenger's head movements. There are small nodding movements of the head *(A)* and largish lateral movements *(B)*. *Lower traces:* acceleration of floor of coach in three directions at 90 mph (from Walsh 1966).

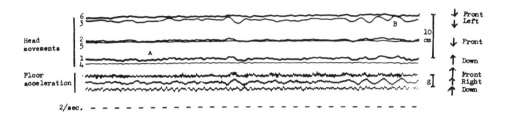

For engine drivers and firemen the situation was worse; Duchesne described a *'maladie des mécaniciens'* (engineers' illness). His work, dating from 1857, was summarised by Schivelbusch (1980). There were generalised, continuous, and persistent pains, accompanied by a feeling of weakness and numbness. It was impossible for them to stand with the feet firmly on the ground. They stood on tiptoe, thereby introducing some springing between their body and the locomotive, or improvised some support from a door mat or a sprung stool.

On British Rail the berths on sleeping carriages are arranged transversely. There was correspondence in the *Times* in December 1985 and January 1986 as to whether this was the best orientation. In Australia and South-West Asia, and in the past in the USA, some coaches have been arranged with bunks arranged longitudinally. The correspondence reached no firm conclusion. Special carriages were once constructed for the transport of invalids (Fig. 11.10).

Because of the oscillations, rail passengers could in the past sometimes be seen propping themselves up in the corner of the carriage to stabilise the head. With aircraft-type seating this is no longer possible (railway engineers please take note).

Head movements have been recorded in passengers travelling by rail. Use was made of beams of light which gave information about the movements of the head in its six degrees of freedom. The apparatus is illustrated in Figure 1.13, but for this application the mirror and lower lamp were to the side of the head. This lamp is referred to as L, that at the vertex as V in Figure 11.11. There were found two principal types of motion (Fig. 11.11). The first, rhythm A, is an oscillation of the head at 3 to 4Hz of up to 4mm amplitude in a rocking motion in the sagittal plane. The head tilts forwards and backwards as it rises and falls. This rhythm was found also on the vertical eye movement records, the eyes being open; the deviations recorded were up to 15°. These oscillations appear to be driven by corresponding

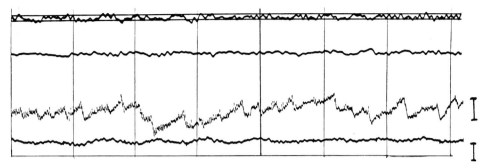

Fig. 11.12. Railway nystagmus. *Top line:* vertical acceleration of floor of coach. *Second line:* lateral acceleration of floor of coach. *Third line:* eye movements, bitemporal electrodes. D.C. recording. *Lowest line:* lateral acceleration of cant rail. *Vertical lines:* seconds. Small calibration bar 0.25g for accelerometers, larger calibration line 5° for eye movements. LMS double bolster.

vibrations of the floor of the coach and are intermittent, only being seen clearly when there is some irregularity of the track, such as when the train is travelling over points. By contrast rhythm B, which was present most of the time, was at 0.5 to 1.5Hz and was a lateral movement of up to 2.8cm amplitude. The record did not consistently show a corresponding lateral acceleration of the floor of the coach. It is likely that tilting of the coach was occurring at a rate to which the trunk and head resonated. This rhythm can very often be seen in railway passengers even on modern trains, but was more conspicuous on vintage trams!

The eyes did not appear to oscillate at the frequency of the B rhythm. It has, however, long been known that moving objects are powerful stimuli to eye movements[3]. When the person looked out of the window rhythmic movements were recorded (Fig. 11.12), a genuine railway nystagmus!

The report quoted above (*Lancet*, 11th January 1862) is again careful and correct as to the types of bodily movement encountered:

> The motion produced in railway travelling is of two kinds—vertical and lateral. The lateral oscillation is more on the narrow than on the broad gauge lines. This may be readily understood on considering the wider base of sustenation on the broad gauge which necessarily diminishes the lateral range of swaying movement. It varies in different carriages in the same train, and according to many conditions. Its extent can scarcely be correctly estimated by personal sensations, as the beautifully elastic apparatus of the body supplies in some measure what the railway carriage is deficient in. But the shadow of an object cast by the flame of a railway lamp, (or of one taken for the purpose), when the flame is steady, will demonstrate how continuous is this influence on passengers. The vertical movement is only occasionally felt. It varies according to the weight of the carriages, and can take place to an extent of three inches and a half without lifting the wheel from the rail.

Vibration induced in the laboratory

The motion of the carriage may be mimicked by the use of a platform tilted

---

[3]Jan Purkinje (1787–1869), Czech physiologist, seems to have been the first to notice rhythmic movements in response to passing images. In 1825 he described nystagmus in the eyes of people watching a procession of cavalry.

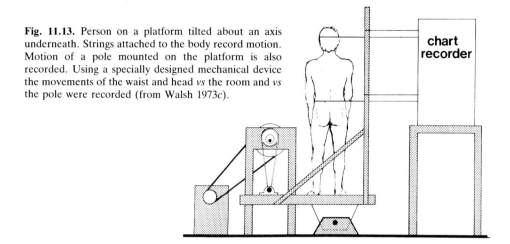

Fig. 11.13. Person on a platform tilted about an axis underneath. Strings attached to the body record motion. Motion of a pole mounted on the platform is also recorded. Using a specially designed mechanical device the movements of the waist and head *vs* the room and *vs* the pole were recorded (from Walsh 1973*c*).

rhythmically through a small angle about an axis below the person (Fig. 11.13). A variable speed motor drove a device in which two eccentric sheaves were used, one inside the other. By locking these together at different angles the throw for each rotation could be varied to obtain varying degrees of tilt. The device was connected through a connecting rod to a pivot on the platform (Walsh 1968*b*). If the platform is driven to rock at different speeds it can be seen that, according to the frequency used, different parts of the body oscillate maximally. It is clear that the body can resonate in a wide variety of different ways. At some frequencies the elbows, at others the knees, at others the head and spine and so forth. With the person seated at the lowest frequency that was used, 0.25Hz, the body moved as a whole. At 1 to 1.5Hz the upper part of the trunk and the head showed substantial oscillations. At higher frequencies the head was relatively well stabilised and the motion was absorbed by the buttocks and the thighs, and at some rates there were striking oscillations of the arms. A reasonably comprehensive description of the multitudinous phenomena would be a complex study. The coupling between different parts of the body will depend to a large extent on the tension in the muscles.

The same apparatus was used to study resonances in the standing position. At least for low frequencies, stabilisation of the head is clearly poorer when the eyes are shut.

Models of the body have been made by engineers interested in the effect of vibration. They show springs, massses and dampers, but those that the author has come across are such gross oversimplifications as to be almost useless (Fig. 11.14).

For the observations in the standing and sitting positions it was informative to record not only the movements of the trunk and head relative to the room but also to record the motion relative to a pole attached to the platform at the same height. A mechanical recorder was employed, strings being used to drive pens on moving paper. By a pantographic system the appropriate subtractions were performed and differential traces obtained by the apparatus which is illustrated elsewhere (Walsh 1973*c*).

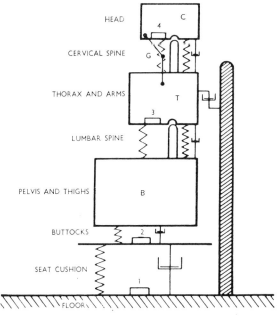

Fig. 11.14. System of masses, pivots, springs and dampers representing a seated man driving a vehicle. The resonant properties of the body are relevant to the design of seats, where there is a substantial vibratory input, such as those of tractors. A model of a man in the standing position based on the work of Coerman is illustrated in Brock (1980). In some coaches the driver's seat is separately sprung. (From Wisner *et al.* 1964, by permission.)

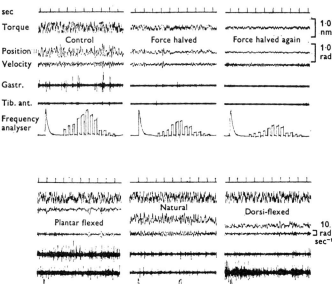

Fig. 11.15. The printed motor is supplied with 'pink' noise, random signals with the lower frequencies somewhat exaggerated. The observations were made at the ankle. *Upper:* effect of reducing force. Gain of velocity channel automatically adjusted so that if force is halved gain is doubled. *Lower:* effect of changing position of foot—considerable EMG activity. Frequency analyser pips from left to right correspond with the following frequencies (Hz):— 2.0, 2.5, 3.1, 3.8, 4.6, 5.7, 7.0, 8.7, 10.7, 13.2, 16.2 and 20.0. The large exponentially decaying pulse indicated dynamic limit of analyser (from Walsh 1973*b*).

TABLE 11.I
Resonant frequencies for different joints in relaxed subjects for motion of moderate amplitude

Part of body	Resonant frequency (Hz)
Hip-abduction/adduction	0.5
Hip-axial rotation of leg	4
Knee	0.5
Ankle	5
Elbow	0.5
Wrist	2.5
Finger	20

Whole body vibration has been studied by many engineers and psychologists. The man has been seated on what is effectively a giant moving coil loudspeaker energised at very low frequencies. The knee jerk is commonly suppressed. Woods (1967) produced vibrations by a hydraulic system. For vertical motion the seated subject showed a body resonance at about 4.5Hz which involved shoulders, arms and legs. For lateral motion the resonance was at 1.5Hz.

At certain frequencies vision is poor. The problem of sight was studied by Huddleston (1970), the optical image being vibrated between 1 and 10Hz, the subject remaining still. At the lower frequencies the eyes automatically followed the movements. In the higher range two virtual images are available as at the turning points of the excursions the velocity of the image drops to zero. The decrement in performance of a visual task was greatest in the intermediate region of the spectrum (3 to 5Hz) where the change over from one strategy to the other had not occurred.

Effects of vibrational 'noise'

In trains, tractors and helicopters the input to the body will consist of a continuous spectrum of vibration, although there will be peaks. An interesting variant on the observations recorded in previous chapters is to excite a printed motor with random electrical signals and observe the resultant motion and electromyographic activity.

A limited number of observations were made at the ankle (Fig. 11.15); the procedure is an alternative method of assessing limb resonances. The resultant shaking of the limb has a spectrum which reflects the relevant biomechanical properties. Held naturally the movement, if the forces are moderately large, possesses a substantial proportion of low-frequency components. These become less conspicuous as the force is reduced. If the foot is plantarflexed or dorsiflexed the motion becomes less and again the fall-off is mostly in the low frequencies. It is less easy to relax when subjected to vibratory noise than to sinusoidal forces.

Comparison of resonant frequencies in different parts of the body

In previous chapters, resonances for a limited number of the possible degrees of freedom of joints of the body have been discussed (Table 11.I).

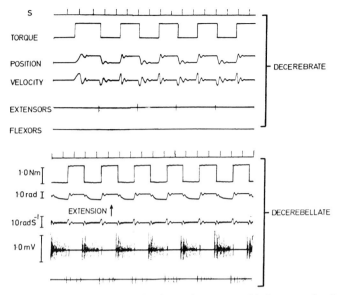

Fig. 11.16. Effect of abruptly alternating torque. Limb moves in direction of applied force until equilibrium is reached after about two cycles of oscillation. This sheep showed no definite stiffness in the decerebrate state and the EMG was almost silent. Following decerebellation the limb moved less for the same force and there was clear stretch activity in the flexor musculature (from Walsh 1977).

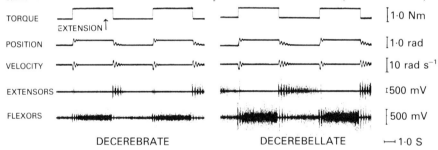

Fig. 11.17. *Left:* decerebrate sheep, tonic stretch activity in flexors, subclonic beats when extensors stretched. *Right:* animal now decerebellate, tonic stretch activity in flexors is enhanced, and now also present in extensors.

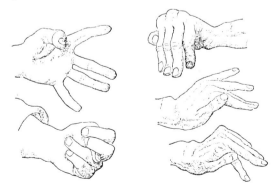

Fig. 11.18. *Left:* athetosis following mild hemiplegia. The hand was in continuous movement between the two positions shown. *Right:* another patient with a similar history (from Gowers 1886).

It is not difficult to see why the effects of whole body vibration vary with frequency. These figures refer to relaxed subjects and moderately large movements. For small movements, or when there is muscular contraction, large elevations can occur as described earlier.

Studies of cerebellar function
In man, lesions of the cerebellum give rise to hyptonia (pp. 116–117). In the pigeon and the cat, however, lesions of the cerebellum heighten the rigidity of the decerebrate animal. Different parts of the cerebellum have developed to different degrees in different species. While the rigidity of the decerebrate animal is due to stretch reflex activity, 'gamma rigidity', that of the decerebrate decerebellate animal has been thought to be due to 'alpha rigidity'. The cerebellar lesion has been believed to inactivate the stretch reflex at the same time as it increases the drive to the motoneurons (Granit et al. 1955). The procedures described in Chapter 4 seemed admirably suited to a reinvestigation of this problem.

Sheep were anaesthetised with halothane and ether and decerebrated using a trephine technique with division of the midbrain, the forebrain and later the cerebellum being removed by suction. The carpal joint was arranged to be concentric with the spindle of a printed motor. The use of sinusoidal torques showed as expected the development of stiffness when the cerebellum was destroyed. A figure illustrating this has been published (Walsh 1977). The degree of stiffness of decerebrate sheep is variable: some were not stiff but stretch reflex activity, elicited by the use of low frequency abruptly alternating torques, became apparent on decerebellation (Fig. 11.16). An example in which there was EMG activity when decerebrate is shown in Figure 11.17. *The stretch reflex increased following decerebellation*.

Spasticity is not relieved by stereotaxic thalamotomy, but dentatomy has its advocates and may possibly be of value also in dystonia (Gornall et al. 1975). Observations were made in conjunction with Professor E. Hitchcock on the effects following dentatomy for children with cerebral palsy. Some had athetosis (Figs 11.18, 11.19). It will be seen that the resonant frequency at the wrist became lower after the operation.

Wings of birds as mechanical oscillators
Greenewalt (1960) used high-speed cine photography to measure the wing-beat frequency of humming birds under such diverse circumstances as hovering, flight at up to 30mph, take-off from a perch and escape behaviour. The frequency was almost constant at about 50Hz under all these conditions. The energy used in powering wings is naturally least when the muscular driving force is in register with the natural resonance of the underdamped system. Other observations illustrate the basic principle. Pelicans flying close to water almost glide, the amplitude of the wing motion becomes very small but the wing frequency is unchanged.

Under unusual and strenuous flight the frequency may vary. Thus a female

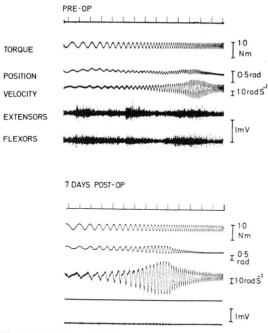

Fig. 11.19. Athetosis. Effect of dentatotomy. The slow variations of the EMG activity are characteristic of this condition and are associated with the continuous shifts of posture. Another name for the condition is 'mobile spasm'. The initial resonant frequency is elevated. After the intervention this drops and the EMG becomes quiescent. Boy aged 12 years, left wrist.

tropical bird, braking before landing at the nesting burrow or eluding a demanding male, changes the frequency for aerodynamic reasons: but the alteration cannot be sustained for long.

Consequences of torsional, inertial and elastic loading
Observations have been made in which a person was called upon to move a finger a certain distance as displayed by a spot on an oscilloscope. Some impediment to the movement was provided by a torque from a printed motor.

In the first experiments the finger was initially held against a stop and thus the magnitude of the resistance was not initially known to the subject. After a number of trials at a torque of 0.17N m there was a reduction to 0.04N m (Desmedt and Godaux 1979).

When the person made the movement as rapidly as possible—a 'ballistic movement'—there was an EMG burst which started before there was detectable motion and lasted 50 to 100ms. The EMG burst was completed by the time the displacement was only half-way there; the remainder of the travel was due to the inertia of the moving part. With the unannounced reduction in the opposing torque, the EMG was not modified in any way but the finger moved further as there was less resistance.

The situation was quite different when the movements were somewhat slower, the person reaching the target in about 0.5s. Once the movement had started, if the opposing torque was unexpectedly reduced, the finger suddenly accelerated; but as the EMG became silent after a latency of 50ms, relaxation occurred earlier. Feedback control is evidently important during most movements when there is an unforeseen alteration of conditions.

Added inertia can increase stability. The balancing pole held by the tightrope walker slows the instabilities and gives him time to make corrective movements. This no doubt requires much learning. When the limbs work into a load which is changed unexpectedly there are naturally at first disturbances of function, as was noticed by Johns and Draper (1964) who studied movements in a tracking task. The person had to move a pointer to a target of one of a number of bulbs. These lit up one at a time, the sequencing being unpredictable. An error signal was obtained when there was a mismatch between the position of the pointer and the target. Part of the error was due to the lag consequent on the reaction time, and it depended on the velocity with which the movement was made.

Normal subjects were able to alter their pattern of control so that large increases of inertial load had almost no effect on performance although when first encountered the velocity of movement was reduced and there was overshooting about the target. When the extra inertia was silently removed, at first the movement was in a series of brief steps; the pointer was not carried on toward the target but would stop short when the muscular activity faded.

Apparently the subject had become accustomed to applying a brief, intense torque to accelerate the inertia followed by a retarding action to prevent overshooting. Connell (1981) wrote of a phenomenon that must have affected millions of army trainees over the years and which is in accord with these experiments:

> During the first few days of riding in a jeep while wearing a steel helmet one's head tends to follow the heavy helmet's inertial path during rapid acceleration and deceleration. That an adaptation has taken place becomes amusingly evident only when one then switches back to lighter headgear, in my case the featherweight overseas cap. For the first few rides you find your head involuntarily nodding at takeoff and tilting backwards during braking periods.

Some limb movements have an important effect in varying inertia. When a skater goes into a spin s/he may start with arms out and then achieve a spectacular increase in the rate of rotation by bringing the arms to the side. Again for a good show much practice will be necessary. The physics of ballet has been studied by Walker (1982). In *jeté en tournant* the dancer leaps into the air with an inconspicuous rotory motion. As she rises she pulls in her arms and brings her legs together; the consequent decrease in her moment of inertia can strikingly increase her rate of spinning.

Some loads that may be encountered give instability. Thus if the wrist is restrained by a fairly soft spring (*e.g.* 2.8N/mm) and the person exerts a fairly strong pull (*e.g.* 100 to 160N), there is likely to develop a vigorous tremor. What appears to happen is that the joint is tuned by the spring to a resonant frequency at which the timing of the stretch reflex gives rise to self re-excitation (Matthews and Muir 1980). Consideration of equation 12 (p. 75) shows that if either stiffness or inertia is increased the system will become more resonant, and thus more prone to oscillate with perturbations.

Most normal people probably develop ankle clonus at times. Sitting in a ramped lecture theatre, with the arch of the foot resting on a narrow support in front, an involuntary oscillation of the musculature of the calf may ensue. The position is inherently unstable; there is liable to be a 'toggle' action at the ankle with the foot moving readily into flexion or extension. Presumably any potential instability caused by neural circuitry is enhanced.

A rock climber told the author that he sometimes developed ankle clonus when standing on tiptoe on a ledge to search for a new hold with his hands. The calf musculature under these circumstances would be supporting the weight of the body and, of course, the moment of inertia of the body about the ankle joint is very large indeed. On lifting a heavy weight there is sometimes an uncontrollable shaking but bearing in mind the wide range of circumstances under which loads may be encountered it is amazing that occurrences of this type are so unusual.

A remarkable achievement of the nervous system is to bring control to a multi-jointed system with an innumerable array of degrees of freedom and so much potential instability.

A resonance theory of posture

When someone stiffens their wrist the resonant frequency may rise from 2 to 10Hz or more, occasionally reaching 16Hz. This latter figure corresponds to an increase of stiffness of 64 times (equation 8, p. 49). All this can be done at a moment's notice by a healthy person. But people with spasticity, rigidity, dystonia and so forth have this ability impaired and may be left with a sustained elevation of muscle tone; they are stuck somewhere in this continuum of possibilities.

In making a movement we use systems which are resonant. Given an impulse there will be overshoot and oscillation at the end of travel unless some action is arranged to take away these effects. The antagonists may need to spring into action followed abruptly at the correct interval by the prime mover again to prevent the limb springing back—a triphasic response. This response was studied by Hallet *et al.* (1975*a*), who observed a 38-year-old man with pansensory neuropathy. He had little subjective appreciation of touch pain, temperature position or vibration. He had no tendon reflexes, and his limbs were accordingly considered deafferented. His triphasic responses were normal so they had reason to believe such motor responses are preprogrammed.

Another man with severe peripheral sensory neuropathy was studied by Rothwell *et al.* (1982). Motor power was almost unaffected. He could produce a very wide range of preprogrammed finger movements with considerable accuracy, involving complex motor synergies of the hand and forearm muscles. He too showed the normal triphasic EMG pattern when executing rapid movements. In spite of this his hands were almost useless to him. He could not grasp a pen and write, fasten buttons or hold a cup. The concatenation of simple motor patterns into a fully formed motor plan needs afferent feedback to inform the CNS of the success of each programme unit, and to modify any errors in execution before proceeding with the plan.

An eminent Norwegian neuroanatomist suffered a stroke at the age of 62 and described his experiences (Brodal 1973). He knew that under normal conditions, the small delicate movements needed to tie a bow-tie follow each other in the proper sequence almost automatically. After his stroke the appropriate finger movements were difficult to perform with sufficient strength, speed and co-ordination but the main problem was that he had to stop because 'his fingers did not know the next move'. It was the sequencing of acts that seemed to be the main problem.

Mosso (1896*b*) wrote of the transformation of voluntary into automatic movements:

> When we first try to execute a series of complicated movements the brain must work hard. If the cells of the upper story—that is, of the convolutions—do not take part in it, all comes to nothing; the assistance of all the organs of sense is necessary in order to shed light on the confusion of orders and counter-orders which must be sent to all the fibres of the muscles. The work is accomplished under the direction of a competent, enlightened administration; but through repetition of the same work, easier paths, broader lines of communication are formed in the lower story of the brain, and gradually the same work can be performed by the cells of the lower part—that is, without the co-operation of the will . . . The oftener a thing is repeated, the more the mechanism tends to become permanent, and it ends in the work being despatched by the less noble parts of the brain.

Bernstein (1967) commented that:

> Like every other form of nervous activity which is structured to meet particular situations, motor co-ordination develops slowly as a result of experiment and exercise. Since co-ordination is, as we have established, a means of overcoming peripheral indeterminacy, it is clear that the basic difficulties for co-ordination consist precisely in the extreme abundance of degrees of freedom, with which the centre is not at first in a position to deal. And, in reality, we observe as a rule that improvement in co-ordination is achieved by utilizing all possible roundabout methods in order to reduce the number of degrees of freedom at the periphery to a minimum. When someone who is a novice at a sport, at playing a musical instrument, or at an industrial process first attempts to master new co-ordination, he is rigidly, spastically fixed and holds the limb involved, or even his whole body, in such a way as to reduce the number of kinematic degrees of freedom which he is required to control.

The subject interested Jones and Round (1990):

> For a task such as reaching out to pick up a cup the simplest sequence of muscle commands must include:
> 1. Activate elbow extensors—arm extends
> 2. Activate finger extensors—hand opens
> 3. Activate finger flexors—cup is grasped
> 4. Activate elbow flexors—cup is brought up towards body.

For a very young child this represent a major achievement requiring great concentration to activate the right muscles in the correct sequence. With time and practice the conscious command 'pick up that cup' serves to initiate the correct sequence of movements without conscious effort.

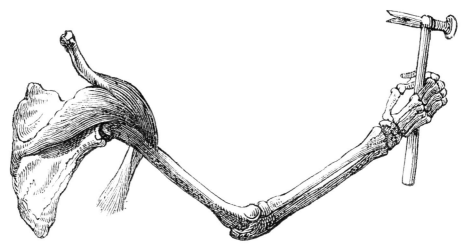

Fig. 11.20. In moving a limb it is necessary for the muscles to control the various jointed segments, and, in using a hammer the task is complicated by the high inertial load. A carpenter, before winding back to strike in earnest, may make two or three light taps to 'line it up'. This is no doubt a method by which the nervous system can select an appropriate programme for motor control (from Bell 1833).

This problem may be thought of in terms of robotics or computer control in which the sequence 1–4 above would be stored as a subroutine to be called up whenever the appropriate movement is required. Nearly every action in the daily life of an adult is encoded in this way and it is only when we encounter an unfamiliar activity such as writing with our non-dominant hand, or learning to ski or to ride a bicycle as an adult that we become conscious of the complex learning process involved.

A trained musician may have remarkable voluntary control. Sir James Paget observed a presto by Mendelssohn being played by a pianist. There were 5595 notes in 4min 3s and it was estimated that there were 72 bimanual finger notes each second.

The changes in the mammalian CNS associated with learning can only be guessed but there was a summary of interesting findings in birds in the *Lancet* (1982):

> The male canary may offer just such a model. The brain nuclei that direct his elaborate song have been identified. He learns his song annually by improvisation and by listening to other canaries, and the more elaborate the song he achieves, the bigger the nuclei. Even his mate, normally songless, can be made to learn to sing pretty well by ovariectomy followed by testosterone treatment; and as she learns, her nuclei enlarge. The neurons in them extend their dendrites to a wider reach. Then in late summer the male's song falls silent and his nuclei shrink, until in the next year the cycle recurs . . . Retracting dendrites (and perhaps vanishing neurons) in the canary losing his song offer a mechanism for loss of a human skill such as writing.

Much, perhaps most, motor behaviour is largely based on prediction. If a person tracks a spot which is moving rhythmically, and which then suddenly disappears for a period, he is likely for some time to keep his movements in register with its presumed course. With the possible exception of fine tuning, the tracking

TABLE 11.II

Fifth centile values for four tests
(Gubbay 1978)

Age (yrs)	Test			
	1. Claps	2. Rolling ball	3. Beads	4. Shapes
6	0	>60	55	>60
7	0–1	>60	55	38
8	1	29	47	36
9	2	28	43	31
10	2	28	39	21
11	2	20	39	21
12	3	18	34	21

The number of claps is given for test 1; the time (s) is given for tests 2 to 4.

does not depend on instant-by-instant feedback. One of the problems in Parkinsonism is this loss of the ability to anticipate what is needed to be done. Patients with this condition may be no better at tracking predictable than unpredictable target movements; they tend to track sine waves as if they were unpredictable irregular movements (Flowers 1978).

In everyday life the resonant frequency of the joints must be continually adapted to the needs of the moment (Fig. 11.20). Furthermore the excitation of muscle must be arranged according to information available as to the characteristics of the system involved. Most movements include more than a single joint. The body is resonant at a series of junctions and unless appropriate control systems were available, unpleasant interactions would build up between one segment and another (pp. 63, 65). Much has to be learnt in early childhood. Teachers and parents are aware that some children are abnormally clumsy. Usually there is no disease of the nervous system. Four tests for clumsiness in children were described by Gubbay (1978):

1. Throwing a tennis-ball into the air and clapping hands (up to four times) before catching it again.
2. Rolling a tennis-ball underfoot in a zig-zag pattern between six matchboxes lined up 30cm apart (timed).
3. Threading 10 beads of 3cm diameter and 0.8cm bore (timed).
4. Inserting six differently shaped objects into appropriate slots (timed).

The publication gives values below which the performance is regarded as substandard at different ages (Table 11.II).

Although the cerebellum has been likened to a computer, it is not always clear what is being computed. I propose that the cerebellum and basal ganglia are concerned with evaluating limb resonances, making adjustments to the frequencies and preparing muscle patterns to minimise the deficiencies in operating movements with systems of this type, or when necessary bringing in enough power to overcome inertia and other resistances to movement. Cerebellar patients cannot appropriately bring muscle activation into play for the proper execution of the triphasic response and there is accordingly hypermetria or hypometria on attempting to reach a target (Hallet *et al.* 1975*b*).

The cerebellum would appear to be an adaptive control system computing information about limb resonances and the best way of achieving motion with minimal wobble. As has been seen, the resonances are much higher for small movements compared with large. Finally, when the range of motion reaches the anatomical limits, the system must become stiff again. Where there is disease of the basal ganglia, as in Parkinsonism, the associated movements are lost or reduced; the arms do not swing during walking. In Huntington's chorea, another disease of the basal ganglia, associated movements are exaggerated contrariwise.

The nervous computations needed must take into account these various factors: the effects of varying load, as when something is carried in the hand, and time-dependent stiffness of a thixotropic type. Complex computations are essential. As the limb stiffens the resonant frequency rises. The ataxia of cerebellar disease is, I suggest, the result of failure to match muscle action to the resonant properties of the joints, and here too a fundamental role must be played by the tone adjusting mechanisms of the brainstem and basal ganglia.

GLOSSARY OF BIOMECHANICAL TERMS

Virtually all the terms in this glossary are used in the book, but the definitions may also be useful for understanding other works where biomechanical problems are discussed. The language of biodynamics is bedevilled for the uninitiated because a number of terms, which are common and flexibly used in everyday English, have precise and rigid definitions in this subject. Some psychological concepts using mechanical linguistic analogies are discussed in Chapter 1. Additional cross references may be found by consulting the index.

Accelerometer
A device commonly fixed to a finger to measure physiological or pathological tremor. Some accelerometers use pressure-sensitive (piezo-electric) crystals, others have elements the electrical resistance of which changes when they are distorted by acceleration. The pressure on the crystal or the distortion of the resistive elements is caused by the transducer, which incorporates a small mass. Owing to its inertia a force is generated corresponding to acceleration (see pp. 34–35). The commonly used accelerometers are sensitive to motion in a straight line—they are linear accelerometers and are inevitably sensitive to tilt. Accelerometers are manufactured which are in general more bulky, and sensitive to rotatory motion only. One device of this type has been used by the author to investigate wrist tremor (see Figs 7.9, 7.21 and 10.13).

Admittance (mobility)
The ratio of velocity to force in an oscillating system. The admittance is the reciprocal of impedance (see 'Impedance' and Fig. 4.10).

Averager
Perhaps the earliest use of signal averaging to enhance the signal/noise ratio was the system used by the 19th-century Scottish physiologist M'Kendrick who wanted to demonstrate the electroretinogram to an audience. He wired six students together and, with their 12 eyes in series, coupled the circuit into his galvanometer. A flash of light gave him the deflection he wanted and which he could not get by the use of one eye alone.

In present-day instruments the averager is triggered by a signal, such as an electrical stimulus, and the currents evoked elsewhere, as in the EEG are recorded and averaged over a number of 'epochs'. Examples of the use of an averager are the observations on tremor in which the component related to the heartbeat has been extracted by triggering the instrument with a signal from the electrocardiogram (see Figs 7.22, 7.23, 7.24).

Ballistic (sclerometer, galvanometer)

Of or pertaining to the throwing of missiles. (O.E.D.)

Instruments in which a relatively heavy mass is free to move with little or no restraining forces (see 'Sclerometer' and Fig. 2.1).

Beats

When sinusoidal forces at two frequencies which are almost the same influence one object, such as a spring-mass system, a characteristic motion known as beating can be observed. As the phase relationships between the two slowly change they are at times acting in the same direction, giving maximal motion, and at times are in opposition giving minimal movement. If the strengths are identical, at this time there will be zero motion as they will exactly cancel one another's effects. There is thus a periodic waxing and waning of amplitude as the relationship shifts between being in and out of phase. The frequency of this beating is equal to the difference in the frequencies of the two forces (see 'Entrain', p. 150 and Figs 10.11, 10.25 and 10.30).

Chirp

This refers to a rhythmic force or torque which increases in frequency. The name originates because if the frequency is high enough a sound may be heard that resembles the chirp of a bird. In biomechanics the frequencies used, at least in the work described in this book, are so slow that the human ear is quite insensitive and the motion is silent. It is customary with chirps to arrange that the frequency rises exponentially. Thus the frequency will double, double again and so on in equal periods of time. The time for the frequency to double is described in terms of octaves per second, thus a value of 0.1 octave s^{-1} means that the frequency will double in 10s. The use of a chirp allows the motion of a limb in response to different rates of pulling to be investigated and is particularly useful in measurements of resonant frequency. Because of the relatively heavy damping of human limbs it is acceptable and convenient for the frequency to be 'swept' relatively rapidly. Slower rates than have been used do not increase accuracy and it is more difficult for the person to remain relaxed for the lengthened period of the chirp. For more resonant systems (*e.g.* steel springs), much slower rates of sweeping are often needed to obtain reliable results (see for instance Fig. 4.6).

Compliance

The reciprocal of stiffness. The more compliant a muscle is, the easier it is to stretch (see 'Stiffness').

Critical damping

With very high and very low levels of damping the new equilibrium position, reached after the application of a force, is achieved only after a long time. The level of damping which gives the most rapid achievement of the new position is known as critical damping and is often approximately the situation found for human limbs (see 'Damping' and pp. 118–119).

Damp(ing)

> To stop the oscillations of a magnetic needle by placing a mass of conducting metal near it. (O.E.D.)

A system which had mass and elasticity alone would vibrate for ever if once disturbed. That this does not occur is due to frictional forces of one or another type. Damping is a form of friction and in the context of this book may be of the nature of viscosity. Engineers often need to introduce damping into systems to prevent unwanted oscillations. One well-known example is that of the shock absorbers ('dampers') of a car. A vehicle in which these components are defective bounces several times after going over a bump in the road. The level of damping is very important in understanding limb control.

Nikola Tesla (1856–1943) attached a variable-frequency, compressed-air-driven vibrator to one of the steel frame members of his laboratory. Unknown to the experimenter there was wholesale breaking of glass, plaster and plumbing in the surrounding buildings as the vibrator passed through its range of frequencies. A 'damping effect' was provided by the arrival of police! (Burton 1968).

The damping coefficient (c) is the torque generated by damping for each unit of rotational velocity to which the system is subjected. The units are accordingly Newton metres per radian per second or N m/rads s^{-1} (see 'Critical damping', 'Q', 'Viscous' and pp. 57–58).

Displacement

Usually used in this book to denote the angular deviation from a zero position, normally the position of rest which the undisturbed limb adopts naturally, or rather, about which it swings during a period of rhythmic excitation by applied torques.

Drag

> The force resisting the motion of a body through a gas or liquid; esp. the resistance along the line of flight to the motion of an aircraft, etc. (O.E.D.)

The word is used by engineers to denote a force opposing motion and dependent on the velocity being of the nature of viscosity (see 'Viscous').

Driving force

An expression coined in connection with observations on postural thixotropy. The limb is set into motion by a rhythmic force, normally of relatively small value: the 'driving force'. After a time adequate for observations on the resultant motion, a perturbation, the 'stirring' force, is interjected, after which the driving force is continued as before. Increased motion, persisting after the disturbance, is suggestive of thixotropy. Confirmation is obtained if the motion is reduced to the initial level when the driving force is cut for a period and then reapplied (see 'Stirring force' and 'Thixotropy', and Figs 7.2, 7.3, 7.5, 7.12, 7.13 and 7.20).

Entrain

> To drag away with or after oneself (O.E.D.)

Entrainement may occur when one oscillatory system influences another. The two systems may synchronise, and at times the transference of quite small amounts of energy from one oscillator to another is adequate for this to occcur. A study of entrainement in the postural system may possibly shed light on the mechanisms of tremor and clonus (see 'Beats', pp. 140–141, p. 148 and pp. 150–151).

Feedback

> The return of a fraction of the output signal from one stage of a circuit, amplifier, etc., to the input of the same or a preceding stage. . . Also a signal so returned. (O.E.D.)

The name given when a signal is routed with some form of amplification back to its source. Biological examples of feedback systems are numerous in physiology, for instance the reflexes, such as the tendon jerks, or the systems regulating breathing and the cardiovascular system (see 'Feedforward', 'Negative position feedback', 'Negative velocity feedback', 'Positive position feedback', 'Positive velocity feedback' and Chapter 5).

Feedforward

A signal routed to an amplifier in anticipation of some imminent event, and, by analogy, physiological signals performing a similar function. Feedforward seems to be very important in the control of limbs, especially when the motion is rapid, as it may be, for instance, in playing the piano. There is no time during the motion for feedback to be effective. Feedforward may be learnt and is perhaps a function of the cerebellum. In the triphasic response an antagonistic muscle is used to check a rapid movement which would otherwise go on for too long and too far.

Filter

A system which freely passes only a certain range of frequencies. A low-pass filter attenuates frequencies above a certain level, a high-pass filter attenuates waveforms below a certain frequency. A band pass filter attenuates above one level and below another. Occasionally used are notch filters which are used to attenuate mains-borne interference being tuned to 50Hz in Europe and 60Hz in the USA. Filters may be mechanical or electrical, but it is usually easier to make them adjustable if they are electrical.

Fourier analysis
> The analysis of a periodic function into a number of simple harmonic functions. (O.E.D.)

When motion repeats itself in equal periods of time it is said to be periodic. The French mathematician J.B.J. Fourier (1768–1830) showed that such periodic functions can be represented by a series of sine and cosine waves (see 'Harmonic').

Fourier's law
> ... that any non-sinusoidal periodic vibration can be regarded as having the sum of a number of sinusoidal vibrations each having a frequency that is an integral multiple of some fundamental frequency. (O.E.D.)

Frequency analysis
See 'Fourier analysis', 'Fourier's law', and Figs 7.22, 7.23, 7.24 and 7.30.

Harmonic
> One of the secondary or subordinate tones produced by the vibration . . . of parts of a sonorous body. (O.E.D.)

An oscillation of a certain frequency may be sinusoidal or some other repetitive waveform. If sinusoidal there is present only the 'fundamental' frequency. If the oscillation is not sinusoidal other components, 'harmonics', are present which may occur at twice, three times and so on of the fundamental frequency. The multiplier is always a whole number. The musical equivalent is of overtones which are responsible for the timbre of an instrument such as the clarinet. Square waves contain the fundamental and the odd-numbered harmonics (p. 61).

Hooke's law
Robert Hooke (1635–1703) did research in a remarkable variety of fields. He was active in astronomy, building an early reflecting telescope and discovering a star in Orion. He also constucted an air pump for Robert Boyle.

> Five years later Hooke discovered the law of elasticity which states that stretching of a solid body (*e.g.* metal, wood) is proportional to the force applied to it. The law laid the basis for studies of stress and strain and for understanding of elastic materials. He applied these studies to the design of a spiral spring for the balance of watches. (New Encyclopaedia Britannica)

The relationship between extension and force for an elastic body such as a spring. If the relationship is linear, Hooke's law is obeyed and often there is some departure (see 'Stiffness', p. 83 and p. 172).

Hysteresis
> ὑστέρησις, a coming short, deficiency . . . to be behind, come late etc . . . Any dependence of the value of a property on the past history of the system to which it pertains. (O.E.D.)

If the relationship between stress and strain is plotted when material is subjected to cyclic load reversal, a loop—the hysteresis loop—is often found. The area of the loop can be interpreted as energy dissipated per cycle. Thus in a spring mass system more work may be done in compressing the spring than is recovered when it is released. When such measurements are made on muscles it is regularly found that there is energy lost during a cycle of stretch and release (see Figs 2.16, 2.17 and 3.3).

Impedance

> Heaviside. Let us call the ratio of the impressed force to the current in a line . . . the impedance of the line . . . Analogous properties of oscillatory mechanical systems. (O.E.D.)

The mechanical impedance of a system is the ratio of the force to the resultant velocity of the motion. As the velocity is highest at resonance, it is at this point that the impedance is least: the restraint is then purely the damping. It may be considered a 'lumped parameter concept', as at other frequencies it will be influenced by inertia and elasticity as well as damping. It has not figured in this work for it has been considered better to analyse the factors one at a time (see also 'Admittance', 'Mobility' and Fig. 4.10).

Inertia

> That property of matter by virtue of which it continues in its existing state, whether of rest or of uniform motion in a straight line, unless that state is altered by an external force. (O.E.D.)

An impediment to rotary motion dependent on the mass of an object and the geometry of the arrangements. The 'moment of inertia' of an object depends on the mass and the square of the distance from the axis of rotation. The units are $kg.m^2$ but often in biomechanics such values are inconveniently large and it is more appropriate to use steps of one-thousandth the size ($g.m^2$).

Inertia is without effect for motion of constant velocity but a resistive torque is generated when the system is accelerated. Inertia is sometimes defined as the 'deadness of matter'. The effects of inertia are familiar to those interested in steam traction engines where the large flywheels smoothed the impulses provided by the cylinders to give an almost uniform rate of rotation. A trapeze artist balancing on a wire uses a long pole which has high inertia to slow his motion and so give him time to make corrective movements.

A human body without inertia is inconceivable outside the realms of science fiction. It would be constantly quivering with the slightest force (see p. 35, Chapter 6, p. 126, and Figs 7.22, 7.26, 7.27, 7.28, 10.29 and 10.30).

Isochronous

> Taking place in or occupying equal times . . . as the vibrations of a pendulum . . . (O.E.D.)

The word refers to the timing of an oscillation when it is found that the time per cycle is the same for large as for small swings. Most lay people expect a large swing to take longer than a small swing, but to a first approximation this is often not true.

Thus a pendulum, provided the swings are not large, will take roughly the same time irrespective of the amplitude (see p. 78).

Isokinetic

> Characterized by no disturbance to the speed and direction of a fluid . . . (O.E.D.)

Literally 'at the same speed'. In some instruments designed to measure tone, motion is imparted by a motor at constant velocity and the resultant force generated is recorded (see pp. 38–40).

In certain commercial devices the person pushes as hard as possible but the velocity attainable is limited to a preset value by adjustments of the controls. The forces that can be generated at different velocities are assessed (see Figs 3.1 and 3.2).

Isometric

> Of equal measure or dimensions (O.E.D.)

A system such as an instrument for testing muscular contractions in which the give is non-existent or negligible. The muscle generates force which is recorded but this is at a predetermined and fixed length. Isometric recordings are normally the method of choice for determining the time course of a contraction (see 'Isotonic' and p. 27).

Isotonic

> Equally stretched, of equal tension or tone. (O.E.D.)

A system such as a freely moving lever attached to a muscle to record the unconstrained motion during a contraction (see 'Isometric' and p. 27).

Kinetic energy

> *Energy*—The power of 'doing work' possessed at any instant by a body or system of bodies. First used by Young . . . to denote what is now called actual, kinetic or motive energy, *i.e.* the power of doing work possessed by a body by virtue of its motion. (O.E.D.)

The kinetic energy of a body is its capacity for doing work insofar as it is in motion. A bullet fired into wood may penetrate a considerable distance: hence before it strikes the target it has the capacity for doing work. The kinetic energy of a rotating body such as a limb in motion is given by the expression $\frac{1}{2}k.v^2$, where 'k' is the moment of inertia and 'v' the angular velocity (see 'Potential energy' and p. 67, footnote).

Linearity (non-linearity)

When a plot of one variable against another is a straight line the relationship is said to be linear. Thus if the amount of stretch is directly proportional to the stretching force Hooke's law is obeyed. If the viscous resistance during a movement is directly proportional to velocity the retarding liquid is Newtonian. If the factors contributing to the operation of a system behave in this way the system itself is linear. Over a certain range the wrist has approximately linear stiffness, and the deviation of the damping from linearity is not very substantial. Thus it is quasilinear. One consequence is that within limits the size of a rhythmic exciting force can be changed without significant alterations of resonant frequency (see 'Damping', 'Hooke's law', 'Resonant frequency', 'Stiffness', 'Damping', and Figs 2.12, 2.13 and 4.18).

Logarithmic decrement
If an oscillatory system is given a tap, a number of oscillatory transients are seen which die away at a rate depending on the damping of the system. The rate of the decay may be such that each swing is a certain fraction of the preceding swing. Such a situation is referred to as a 'logarithmic decay' or 'decrement'; it is found when damping forces are linearly related to the velocity of movement as when they arise from the drag of a viscous fluid having Newton properties (see 'Damping', 'Transients', 'Viscosity' and Figs 7.26, 7.27, 7.28, 7.30, 8.9, 8.10, 8.11 and 8.12).

Magnification factor
This relates to the properties of a resonant system and is the ratio of the amplitude of the motion at resonance to the static deflection caused by a steady application, instead of a rhythmic application, of the force (see 'Q' and p. 57, footnote).

Mobility (admittance)
The mechanical mobility is the ratio of attained velocity to the applied force in a rhythmically driven system. It is hence the reciprocal of the mechanical impedance. At resonance it reaches its highest value, where it is convenient to measure it; it is then inversely related to the damping (see 'Damping', 'Impedance' and Table 6.I).

Momentum
> The 'quantity' of motion of a moving body; the product of mass by the velocity of a body . . . the effect of inertia in the continuance of motion after the impulse has ceased; impetus gained by movement. (O.E.D.)

The change of momentum produced by a force is directly proportional to the magnitude of the force and to the time for which it acts, *i.e.* the 'impulse'.

Myotonograph
An instrument for measuring muscle tone (for an early example see Fig. 2.6).

Natural frequency
The frequency at which a mass spring system would oscillate after a disturbance *if*

damping had negligible effects. In real life there is some influence but this is normally small, and the system resonates at a different rate, the resonant frequency. Even with the relatively high damping of human limbs the actual rate of vibration is not likely to be more than 20 per cent lower than that calculated on the basis of equation 8 (p. 49), the difference for the purposes of the present studies have been neglected (see 'Resonant frequency').

Negative damping

The inherent damping of a limb is due to viscosity and other factors which impede motion according to the velocity. By the use of feedback to a motor coupled to a limb it is possible to impart a torque which acts in the opposite sense, motion being enhanced in the direction in which the limb is travelling (see 'Damping' and 'Positive velocity feedback').

Negative position feedback

By the use of feedback to a motor, a torque can be generated which resists the displacement of a limb from the resting position according to the extent of the deviation. The stiffness of the system is thus increased (see 'Positive position feedback', 'Stiffness' and pp. 69–73).

Negative velocity feedback

If a signal corresponding to velocity is fed to a motor coupled to a limb with appropriate polarity, there is additional resistance (varying according to the velocity) in any active or passive movement (see 'Positive velocity feedback' and p. 59).

Period

> The time during which anything runs its course, time of duration. (O.E.D.)

The reciprocal of frequency. The period of an oscillation is the time taken forr one cycle to be completed (see 'Simple harmonic motion').

Phase

> A particular stage or point in a recurring sequence of movements or changes *e.g.* a vibration or undulation (O.E.D.)

When two oscillations of the same frequency are in step they are said to be 'in phase'. When the motion is in opposite senses they are 'out of phase'. When one oscillation is at its turning point whilst the other is at its point of maximal velocity they are 'in quadrature'. Phase is measured in degrees (0–360) or in radians (0–2π). When an event regularly occurs at a certain point of a cycle it is said to be 'phase-locked'. At resonance, torque and velocity are in phase; below resonance, velocity is phase advanced on torque; above resonance it is phase-retarded. For sine-wave motion, velocity always leads position by 90° ($\frac{\pi}{2}$) and acceleration again leads velocity by 90° ($\frac{\pi}{2}$) (see 'Simple harmonic motion').

Positive position feedback
A signal corresponding to the deviation from the resting position is fed back to a motor coupled to a limb so that a torque is generated accentuating the deviation. The natural elasticity of the limb can thus be partially or wholly cancelled (see Fig. 6.2).

Positive velocity feedback
A signal corresponsing to angular velocity is fed back to a motor coupled to the limb so that, when movement takes place in one direction, a torque is generated which assists the motion. With low gain, voluntary movements show increased instability; with higher gain, the limbs bursts spontaneously into oscillation at the resonant frequency. The circuits then saturate, they reach maximum capacity and the torque is generated in the form of square waves. This system provides a rapid and useful way of assessing stiffness as the resonant frequency is related to the square root of stiffness (see pp. 59–66 and 71–75).

Potential energy
The potential energy of a system is the amount of work which it is capable of performing in virtue of its position.
    In the resonance of a limb the maximum potential energy is when the displacement is maximal and the velocity is zero—*i.e.* at the turning points (see 'Kinetic energy').

Printed motor
A device that converts an electrical current into the corresponding mechanical effect—a torque. Used extensively in this book to investigate resonances of human limbs (see 'Resonance', 'Resonant frequency', 'Torque' and pp. 45–47).

Pseudo-differentiator
An electrical circuit which performs a function approximating to the mathematical differentiation of a signal, as in obtaining a velocity signal from a potentiometer the output of which corresponds to position (see 'Tachometer' and p. 48).

Pulling
A useful term evidently first introduced in a technical sense by Roberts (1963) to indicate that a muscle is subjected to a force, in contradiction to being stretched through a certain defined distance. Pulling has proved to be a much more useful method of analysing muscle tone than stretching (see 'Stretching', p. 38 and Chapter 4).

Q (quality factor)
A measure of the sharpness of tuning. The value may be obtained by dividing the resonant frequency by the difference between the upper and lower frequencies at which the amplitude has dropped to 70 per cent of that at resonance. The greater the damping, the lower the value for Q (see 'Damping', 'Magnification factor', p. 57 and p. 75).

Random forces
With the use of torques which slowly alternate, useful information about the properties of limb control may rapidly be obtained. Some people anticipating the change, however, overreact voluntarily and so upset the observations. In them it may be useful to randomise the time at which the transitions occur. This may be achieved by the use of a 'binary noise generator' (see Fig. 9.8).

Random signals clocked at a rapid rate may be described as being 'white noise' if the frequency components are uniform in the part of the frequency spectrum under consideration, the analogy being with the spectral composition of white light. If the low frequency components are predominant the signal is described as being 'pink noise' or 'red noise' depending on the degree of accentuation. The use of random signals is an alternative method of evaluating resonant frequencies, but has not been extensively used (see however Fig. 11.15).

The inputs to the body may be very wide in frequency terms and essentially random, *e.g.* when travelling in early railway trains and helicopters (see pp. 174–181).

Resonance
The word was first used about sound—Latin *resonantia* ('echo'). Early usage may be illustrated by a quotation from Burney's *General History of Music:*

> Resonance is but an aggregate of echos or of quick repetitions and returns of the same sound. (O.E.D.)

Piano or guitar strings provide good examples of resonant systems. Struck or plucked they vibrate at a rate dependent on their mass and stiffness. The higher the mass, the lower the frequency; the higher the stiffness, the higher the frequency.

The term is now much more widely used by both mechanical and electrical engineers. Most objects have mass and elasticity and will show the phenomenon of resonance if the frictional forces, such as viscosity are below a certain level. For rotary systems such as a part of a limb moving around the proximal joint, the relationship is between inertia and rotary stiffness. Resonance occurs when the effects are equal and opposite (see 'Damping', 'Resonant frequency' and pp. 48–52).

Resonant frequency
When a resonant system is excited by a rhythmic force, the maximal motion occurs at the resonant frequency. If the inertia (for instance of a limb) can be considered constant, the resonant frequency reflects the stiffness of the system (see 'Resonance' and equation 8, p. 49).

Sclerometer
> An instrument for measuring the hardness of crystals. (O.E.D.)

An instrument for measuring hardness by indentation (see Fig. 2.1).

Simple harmonic motion (SHM)
> *Harmonic motion*—A periodic motion, which in its simplest form (simple harmonic motion) is like that of a point in a vibrating string, and is identical with the resolved part, parallel to a diameter, of uniform motion in a circle. (O.E.D.)

This may be regarded as the simplest type of rhythmic motion; it is sometimes referred to as being a 'pure sine wave'. The frequency can be specified as so many cycles per second (hertz or Hz), or thinking in terms of the projection of circular motion it may be specified as so many radians per second, there being 2π radians per cycle. Other rhythmic signals may be broken down into components of this type by frequency analysis (see 'Fourier Analysis' above).

Step function
A signal the level of which abruptly changes. One example is the rectangular alternating torques used for the investigations described in Chapter 9.

Step functions are one way of assessing the resonance of a system; an underdamped system rings at the resonant frequency.

Stiffness
In lay use, stiff has a wide variety of meaning. In this book its use is restricted to a technical engineering sense. Stiffness is the relationship between extension and extending force (see 'Hooke's law, 'Linearity', p. 42 and Fig. 2.12).

Stirring force
A term coined for observations on thixotropy. A small rhythmic force is applied to a limb then for a short period a larger force—the 'stirring force'—is interjected. After that the force reverts to the previous lower level (see 'Driving force', 'Thixotropy', and Figs 7.2, 7.3, 7.5, 7.7, 7.12, 7.13 and 7.20).

Strain
The displacement corresponding to a certain force (see 'Hooke's law', 'Stiffness', 'Stress' and p. 38).

Stress
> A force acting on or within a body or structure and tending to deform it; now usu. the intensity of this force per unit area. (O.E.D.)

The force corresponding to a certain displacement (see 'Hooke's law', 'Stiffness', 'Strain' and p. 38).

Stretching
Used in a technical sense by Roberts (1963). The term indicates that the muscle is stretched through a certain distance (see 'Pulling' and Chapter 3).

Tachometer
> An instrument by which the velocity of machines is measured. (O.E.D.)

An instrument which gives an electrical signal corresponding to angular velocity. In many of the observations in this book, however, angular velocity has been obtained by the differentiation of a position signal (see 'Pseudo-differentiator', p. 73 and Fig. 7.29).

Torque
> The twisting or rotary force in a piece of mechanism (as a measurable quantity); the moment of a sytem of forces producing rotation. (O.E.D.)

A rotary force that twists the object. Torques in this work have been generated by printed motors and occasionally by basket-wound motors (see 'Printed Motor' and p. 34).

Transients
After a disturbance a resonant system may 'ring' with a series of oscillations which, with greater or lesser speed, decay. Such ringing is common in the flexor musculature of the arm of hemiplegic patients as a result of stretching.

Ringing contributes substantially to the phenomenon of physiological tremor, as for very small displacements the postural system is very stiff and has a comparatively high resonance (see 'Logarithmic decrement' and Figs 7.9, 7.30, 8.9, 8.10, 8.11 and 8.12).

Vibration absorber
A system where by allowing motion of an appendage the effect of vibration on a main structure is reduced (see p. 63).

Viscous (viscosity)
> Imperfectly fluid; intermediate between solid and fluid; adhesively soft. (O.E.D.)

This property is characteristic of thick fluids such as treacle, but even gases such as air have viscosity—witness aerodynamic drag. Joints, fascial planes and tendon sheaths are lubricated by slimy fluid which has significant viscosity. Viscosity gives rise to opposition to motion related to velocity. With an ideal 'Newtonian' fluid the retarding action is linear but many fluids including synovial fluid do not obey this relationship; they are 'non-Newtonian' (see 'Damping, 'Drag', 'Linearity', pp. 41–42 and Fig. 2.13).

REFERENCES

Agate, F.J., Doshay, L.J. (1956) 'Quantitative measurement of therapy in paralysis agitans.' *Journal of the American Medical Association*, **160**, 352–354.
Alexander, R.M. (1988) *Elastic Mechanisms in Animal Movement.* Cambridge: Cambridge University Press.
—— (1989) 'Muscles for the job.' *New Scientist*, **122**, 50–53.
Alkon, D.L. (1987) *Memory Traces in the Brain.* Cambridge: Cambridge University Press.
André-Thomas, Ajuriaguerra, J. de (1949) *Étude Sémiologique du Tonus Musculaire.* Paris: Flammarion.
—— Chesni, Y., Dargassies, S.-A. (1960) *The Neurological Examination of the Infant. Little Club Clinics in Developmental Medicine, No. 1.* London: Medical Advisory Committee of the National Spastics Society.
Andry, N. (1743) *Orthopaedia; or the Art of Correcting and Preventing Deformities in Children.* London: Millar.
Ashworth, B. (1964) 'Preliminary trial of carisoprodol in multiple sclerosis.' *Practitioner*, **192**, 540–542.
Atha, J., Yeadon, M.R., Sandover, J., Parsons, K.C. (1985) 'The damaging punch.' *British Medical Journal*, **291**, 1756–1757.
Bajd, T., Vodovnik, L. (1984) 'Pendulum testing of spasticity.' *Journal of Biomedical Engineering*, **6**, 9–16.
Barnes, W.J.P., Gladden, M.H. (1985) *Feedback and Motor Control in Invertebrates and Vertebrates.* London: Helm.
Barnett, C.H., Cobbold, A.F. (1969) 'Muscle tension and joint mobility.' *Annals of Rheumatic Diseases*, **28**, 652–654.
Basmajian, J.V. (1957) 'New views on muscular tone and relaxation.' *Canadian Medical Association Journal*, **77**, 203–205.
—— Steko, G. (1963) 'The role of muscles in arch support of the foot.' *Journal of Bone and Joint Surgery*, **45A**, 1184–1190.
Bate-Smith, E.C., Bendall, J.R. (1949) 'Factors determining the time course of rigor mortis.' *Journal of Physiology*, **110**, 47–65.
Beighton, P., Grahame, R., Bird, H. (1989) *Hypermobility of Joints, 2nd Edn.* London: Springer.
Bell, C. (1824) *Essays on the Anatomy and Philosophy of Expression.* London: John Murray.
—— (1833) *The Hand. Its Mechanism and Vital Endowments as Evincing Design.* London: Pickering.
Benson, A.J., Gedye, J.L. (1961) 'Some supraspinal factors influencing generalised muscle activity.' *In* Turnbull, P.C. (Ed.) *Symposium on Skeletal Muscle Spasm.* Leicester: Franklyn, Ward and Wheeler.
Benthuysen, J.L., Smith, N.T., Sandford, T.J., Head, N., Dec-Silver, H. (1986) 'Physiology of Alfentanil-induced rigidity.' *Anesthesiology*, **64**, 440–446.
Berne, R.M., Levy, M.N. (1988) (Eds) *Physiology. 2nd Edn.* St. Louis: Mosby.
Bernstein, N. (1967) *The Co-ordination and Regulation of Movements.* Oxford: Pergamon.
Berthoz, A., Metral, S. (1970) 'Behaviour of a muscular group subjected to a sinusoidal and trapezoidal variation of force.' *Journal of Applied Physiology*, **29**, 378–384.
Bird, H.A., Stowe, J. (1982) 'The wrist.' *Clinics in Rheumatic Diseases*, **8**, 559–569.
Borelli, G.A. (1685) *De Motu Animalium.* Batavis: Lugduni.
Bouisset, S., Pertuzon, E. (1968) 'Experimental determination of the moment of inertia of limb segments.' *Biomechanics*, **1**, 106–109.
Braune, W., Fischer, O. (1892) *Bestimmung der Trägheitsmomente des menschlichen Körpers und seiner Glieder.* Leipzig: Hirzel. (Translated in 1988 by P. Maquet and R. Furlong. Berlin: Springer.)
Brenner, B. (1990) 'Muscle mechanics and biochemical kinetics.' *In* Squire, J.M. (Ed.) *Molecular Mechanisms in Muscular Contraction.* Basingstoke: Macmillan.
Broch, J.T. (1980) *Mechanical Vibrations and Shock Measurements.* Naerum: Brüel and Kjær.
Brodal, A. (1973) 'Self-observations and neuro-anatomical considerations after a stroke.' *Brain*, **96**, 675–694.

Granit, R., Holmgren, B., Meron, P.A. (1955) 'The two routes for excitation of muscle and their subservience to the cerebellum.' *Journal of Physiology*, **130**, 213-224.

Greenewalt, C.H. (1960) 'The wings of insects and birds as mechanical oscillators.' *Proceedings of the American Philosophical Society*, **104**, 605-611.

Gruener, R., McArdle, B., Ryman, B.E., Weller, R.O. (1968) 'Contracture of phosphorylase deficient muscle.' *Journal of Neurology, Neurosurgery and Psychiatry*, **31**, 268-283.

Gubbay, S.S. (1978) 'The management of developmental apraxia.' *Developmental Medicine and Child Neurology*, **20**, 643-646.

Guyton, A.C. (1981) *Textbook of Medical Physiology*. 6th Edn. Philadelphia: Saunders.

Haines, R.W. (1934) 'On muscles of full and of short action.' *Journal of Anatomy*, **69**, 20-24.

Haley, S.M., Inacio, C.A. (1990) 'Evaluation of spasticity and its effect on motor function.' *In* Glenn, M.B., Whyte, J. (Eds) *The Practical Management of Spasticity in Children and Adults.* Philadelphia: Lea and Febiger.

Hallett, M., Shahani, B.T., Young, R.R. (1975a) 'EMG analysis of stereotyped voluntary movements in man.' *Journal of Neurology, Neurosurgery and Psychiatry*, **38**, 1154-1162.

—— —— —— (1975b) 'EMG analysis of patients with cerebellar deficits.' *Journal of Neurology, Neurosurgery and Psychiatry*, **38**, 1163-1169.

Harreveld, A. van, Bogen, J.E. (1961) 'The clinging position of the bulbocapninized cat.' *Experimental Neurology*, **4**, 241-261.

Harris, P., Walsh, E.G. (1972) 'Simultaneous electrical and mechanical recording from postural muscles in a paraplegic patient.' *Paraplegia*, **9**, 229-230.

Henderson, Y. (1938) *Adventures in Respiration*. London: Baillière, Tindall and Cox.

Hildebrand, M. (1960) 'How animals run.' *Scientific American*, **202**, 148-157.

Hill, D.K. (1968) 'Tension due to interaction between the sliding filaments in resting striated muscle. The effect of stimulation.' *Journal of Physiology*, **199**, 637-684.

Hodes, R., Gribetz, I., Hodes, H.L. (1962) 'Abnormal occurrence of the ulnar nerve-hypothenar muscle H-reflex in Sydenham's chorea.' *Pediatrics*, **30**, 49-56.

Hodgkinson, L. (1988) *The Alexander Technique*. London: Piatkus.

Holbourn, A.H.S. (1943) 'Mechanics of head injuries.' *Lancet*, **2**, 438-441.

Holmes, G. (1917) 'The symptoms of acute cerebellar lesions due to gun-shot injuries.' *Brain*, **40**, 461-535.

Howell, W.H. (1905) *A Text-book of Physiology*. Philadelphia: W.B. Saunders.

Huddleston, J.H.F. (1970) 'Tracking performance on a visual display apparently vibrating at one to ten Hertz.' *Journal of Applied Psychology*, **54**, 401-408.

Hufschmidt, A., Schwaller, I. (1987) 'Short-range elasticity and resting tension of relaxed human lower leg muscles.' *Journal of Physiology*, **391**, 451-465.

Hunter, J. (1837) 'Croonian lectures on muscular motion.' *In* Palmer, J.F. (Ed.) *The Works of John Hunter, Vol. IV*. London: Longman.

Huxley, H.E., Brown, W. (1967) 'The low-angle X-ray diagram of vertebrate striated muscle and its behaviour during contraction and rigor.' *Journal of Molecular Biology*, **30**, 383-434.

Iggo, A., Walsh, E.G. (1960) 'Selective block of small fibres in the spinal roots by phenol.' *Brain*, **83**, 701-708.

Jain, S.K., Subramanian, S., Julka, D.B., Guz, A. (1972) 'Search for evidence of lung chemoreceptors in man: a study of respiratory and circulatory effects of phenyldiguanide and lobeline.' *Clinical Science*, **42**, 163-177.

Johns, R.J., Draper, I.T. (1964) 'The control of movement in normal subjects.' *Bulletin of the Johns Hopkins Hospital*, **115**, 447-464.

Jones, D.A., Round, J.M. (1990) *Skeletal Muscle in Health and Disease*. Manchester: Manchester University Press.

Kasdon, D.L., Abramovitz, J.N. (1990) 'Neurosurgical approaches.' *In* Glenn, M.B., Whyte, J. (Eds) *The Practical Management of Spasticity in Children and Adults.* Philadelphia: Lea and Febiger.

Keith, A. (1919) *Menders of the Maimed*. London: Frowde.

Kerr, J.D.O., Scott, L.D.W. (1936) 'The measurement of muscle tonus.' *British Medical Journal*, **2**, 758-760.

Kleist, K. (1927) 'Gegenhalten (motorischer Negativismus), Zwangsgreifen und Thalamus opticus.' *Monatsschrift für Psychiatrie und Neurologie*, **65**, 317-396.

Knights, D.E. (1975) 'Prospects for the printed motor.' *Engineering*, **215**, 199–202.
Knutsson, E. (1985) 'Quantification of spasticity.' *In* Struppler, A., Weindl, A. (Eds) *Electromyography and Evoked Potentials*. Berlin: Springer.
Kok, O., Bruyn, G.W. (1962) 'An unidentified hereditary disease.' *Lancet*, **1**, 1359.
Krott, H.M., Jacobi, H.M. (1972) 'Neurophysiologische Untersuchungen der Eigenreflexstörung beim Adie-Syndrome.' *Archiv für Psychiatrie und Nervenkrankheiten*, **215**, 338–361.
Kugelberg, E. (1946) 'Injury activity and trigger zones in human nerves.' *Brain*, **69**, 385–396.
Kushida, C.A., Baker, T.L., Dement, W. (1985) 'Electroencephalographic correlates of cataplectic attacks in narcoleptic canines.' *Electroencephalography and Clinical Neurophysiology*, **61**, 61–70.
Laarse, W.D. van der, Oosterveld, W.J. (1971) 'Muscle tonus measurements.' *Biomechanics*, **2**, 308–315.
Lakie, M. (1981) 'An investigation into muscle tone using printed motors as torque generators.' University of Edinburgh: Ph.D. thesis.
—— Walsh, E.G., Wright, G.W. (1979a) 'Wrist compliance.' *Journal of Physiology*, **295**, 98–99P.
—— —— —— (1979b) 'Cooling and wrist compliance.' *Journal of Physiology*, **296**, 47–48P.
—— Tsementzis, S.T., Walsh, E.G., Wright, G.W. (1980a) 'Anaesthesia does not (and cannot) reduce muscle tone?' *Journal of Physiology*, **301**, 32P.
—— Walsh, E.G., Wright, G.W. (1980b) 'Thixotropy—a general property of the postural system.' *Journal of Physiology*, **305**, 73–74P.
—— —— (1981) 'Sudden increase of wrist stiffness for small movements—demonstrated by positive velocity feedback.' *Journal of Physiology*, **312**, 46–47P.
—— —— —— (1981) 'Measurements of inertia of the hand, and the stiffness of the forearm muscles using resonant frequency methods, with added inertia or position feedback.' *Journal of Physiology*, **310**, 3–4P.
—— —— (1982) 'Cold adiadokokinesia.' *Journal of Physiology*, **328**, 41–42P.
—— Scott, D.B., Walsh, E.G., Wright, G.W. (1983a) 'Resonance at the wrist in anaesthetised subjects.' *Journal of Physiology*, **338**, 32P.
—— Walsh, E.G., Wright, G.W. (1983b) 'On tuning a physiological tremor.' *Journal of Physiology*, **334**, 32–33P.
—— —— —— (1984a) 'Passive wrist movements—thixotropy—measurement of memory time.' *Journal of Physiology*, **346**, 6P.
—— —— —— (1984b) 'Resonance at the wrist demonstrated by the use of a torque motor: an instrumental analysis of muscle tone in man.' *Journal of Physiology*, **353**, 265–285.
—— —— —— (1986a) 'Control and postural thixotropy of the forearm muscles; changes caused by cold.' *Journal of Neurology, Neurosurgery and Psychiatry*, **49**, 69–76.
—— —— —— (1986b) 'Passive mechanical properties of the wrist and physiological tremor.' *Journal of Neurology, Neurosurgery and Psychiatry*, **49**, 669–676.
—— Robson, L.G. (1988a) 'Thixotropy: the effect of stretch size in relaxed frog muscle.' *Quarterly Journal of Experimental Physiology*, **73**, 127–129.
—— —— (1988b) 'Thixotropic changes in human muscle stiffness and the effects of fatigue.' *Quarterly Journal of Experimental Physiology*, **73**, 487–500.
—— Walsh, E.G., Wright, G.W. (1988) 'Assessment of human hemiplegic spasticity by a resonant frequency method.' *Clinical Biomechanics*, **3**, 173–178.
Lambert, E.H., Underdahl, L.O., Beckett, S., Mederos, L.O. (1951) 'A study of the ankle jerk in myxedema.' *Journal of Clinical Endocrinology*, **11**, 1186–1206.
Lancet (1862) 'The influence of railway travelling on public health. Report of the commission ii.' (Editorial.) *Lancet*, **1**, 48–53.
—— (1982) 'Winter's cramp.' (Editorial.) *Lancet*, **2**, 969.
Landsmeer, J.M.F. (1949) 'The anatomy of the dorsal aponeurosis of the human finger and its functional significance.' *Anatomical Record*, **104**, 31–44.
Laufman, H. (1951) 'Profound accidental hypothermia.' *Journal of the American Medical Association*, **147**, 1201–1212.
Lawrence, W. (1866) *Lectures on Comparative Anatomy, Physiology, Zoology and the Natural History of Man*. London: Bell and Daldy.
Lehmann, J. F., Price, R., deLateur, B.J., Hinderer, S., Traynor, C. (1989) 'Spasticity: quantitative measurements as a basis for assessing effectiveness of therapeutic intervention.' *Archives of*

Physical Medicine and Rehabilitation, **70**, 6–15.
Levine, I.M. (1964) 'Symposium on skeletal muscle hypertonia.' *Clinical Pharmacology and Therapeutics*, **5**, 800–966.
—— Jossmann, P.B., deAngelis, V. (1972) 'The quantitative evaluation of CIBA 34,647-Ba. A preliminary report.' *In* Birkmayer, W. (Ed.) *Spasticity—A Topical Survey.* Bern: Huber.
Levy, R. (1963) 'The relative importance of the gastrocnemius and soleus muscles in the ankle jerk of man.' *Journal of Neurology, Neurosurgery and Psychiatry*, **26**, 148–150.
Lipsitz, L.A. (1983) 'The drop attack: a common geriatric symptom.' *Journal of the American Geriatrics Society*, **31**, 617–620.
Lloyd, A.R., Hales, J.P., Gandevia, S.C. (1988) 'Muscle strength, endurance and recovery in the post-infection fatigue syndrome.' *Journal of Neurology, Neurosurgery and Psychiatry*, **51**, 1316–1322.
Luciani, L. (1915) *Human Physiology, Vol. 3. Muscular and Nervous System.* London: Macmillan.
McComas, A.J., Sica, R.E.P., Upton, A.R.M., Aguilera, N. (1973) 'Functional changes in motoneurones of hemiparetic patients.' *Journal of Neurology, Neurosurgery and Psychiatry*, **36**, 183–193.
MacDougall, J.D.B., Andrew, B.L. (1953) 'An electromyographic study of the temporalis and masseter muscles.' *Journal of Anatomy*, **87**, 37–45.
McMahon, T.A. (1975) 'Using body size to understand the structural design of animals: quadrupedal locomotion.' *Journal of Applied Physiology*, **39**, 619–627.
Maloiy, G.M.O., Heglund, N.C., Prager, L.M., Cavagna, G.A., Taylor, C.R. (1986) 'Energetic cost of carrying loads: have African women discovered an economic way? *Nature*, **319**, 668–669.
Marey, E.J. (1874) *Animal Mechanism.* London: King.
—— (1895) *Movement.* London: Heineman.
Marsden, C.D., Foley, T.H., Owen, D.A.L., McAllister, R.G. (1967) 'Peripheral ß-adrenergic receptors concerned with tremor.' *Clinical Science*, **33**, 53–65.
—— Merton, P.A., Morton, H.B. (1976a) 'Servo action in the human thumb.' *Journal of Physiology*, **257**, 1–44.
—— —— —— (1976b) 'Stretch reflex and servo action in a variety of human muscles.' *Journal of Physiology*, **259**, 531–560.
Marshall, L. (1975) *Wake up to Yoga.* London: Ward Lock.
Martindale, W. (1989) *The Extra Pharmacopoeia. 29th Edn.* London: Parmaceutical Press. p. 1444.
Matsen, F.A. (1980) *Compartmental Syndromes.* New York: Grune and Stratton.
Matthews, P.B.C. (1966) 'The reflex excitation of the soleus muscle of the decerebrate cat caused by vibration applied to its tendon.' *Journal of Physiology*, **184**, 450–472.
—— (1972) *Mammalian Muscle Receptors and their Central Actions.* London: Arnold.
—— (1991) 'The human stretch reflex and the motor cortex.' *Trends in Neurosciences*, **14**, 87–91.
—— Muir, R.B. (1980) 'Comparison of electromyogram spectra with force spectra during human elbow tremor.' *Journal of Physiology*, **302**, 427–441.
Maughan, R.J., Watson, J.S., Weir, J. (1984) 'The relative proportions of fat, muscle and bone in the normal human forearm as determined by computed tomography.' *Clinical Science*, **66**, 683–689.
May, P.R.A., Fuster, J.M., Haber, J., Hirschman, A. (1979) 'Woodpecker drilling behaviour. An endorsement of the rotational theory of impact brain injury.' *Archives of Neurology*, **36**, 370–373.
Mitchell, S.W. (1872) *Injuries of Nerves.* (Reprinted 1964.) New York: Dover.
Mizuno, Y., Tanaka, R., Yanagisawa, N. (1971) 'Reciprocal group 1 inhibition of triceps surae motoneurones in man.' *Journal of Neurophysiology*, **34**, 1010–1017.
Mosso, A. (1896a) 'Description d'un Myotonomètre pour étudier la tonicité des muscles chez l'homme.' *Archives Italiennes de Biologie*, **25**, 349–384.
—— (1896b) *Fear.* London: Longmans Green.
—— (1904) *Fatigue.* London: Swan Sonnenschein.
Moyer, H.N. (1911) 'A new diagnostic sign in paralysis agitans: the cogwheel resistance of the extremities.' *Journal of the American Medical Association*, **57**, 2125.
Muir, A.L., Percy-Robb, I.W., Davidson, I.A., Walsh, E.G., Passmore, R. (1970) 'Physiological aspects of the Edinburgh Commonwealth Games.' *Lancet*, **2**, 1125–1128.
Müller, J. (1842) *Elements of Physiology.* London: Taylor and Walker.
Muybridge, E. (1957) *Animals in Motion.* New York: Dover.
Nathan, P.W. (1968) 'Motor effects of differential block of spinal roots in spastic patients.' *Journal of*

the Neurological Sciences, **8**, 9–26.
Newman, D.G., Pearn, J., Barnes, A., Young, C.M., Kehoe, M., Newman, J. (1984) 'Norms for hand grip strength.' *Archives of Disease in Childhood*, **59**, 453–459.
Paintal, A.S. (1986*a*) 'The visceral sensations—some basic mechanisms.' *Progress in Brain Research*, **67**, 3–19.
—— (1986*b*) 'The significance of dry cough, breathlessness and muscle weakness.' *Indian Journal of Tuberculosis*, **33**, 51–55.
—— Walsh, E.G. (1981) 'Inhibition of tonic stretch reflex by J receptor activity.' *Journal of Physiology*, **316**, 22–23P.
Paley, W. (1846) *Natural Theology, or Evidences of the Existence and Attributes of the Deity*. London: Milner.
Parkinson, J. (1817) *An Essay of the Shaking Palsy*. London: Sherwood, Neerly & Jones. Reprinted in Critchley, M. (1955) (Ed.) *James Parkinson (1755–1824)*. London: Macmillan.
Paulus, W., Straube, A., Brandt, T.H. (1987) 'Visual postural performance after loss of somatosensory and vestibular function.' *Journal of Neurology, Neurosurgery and Psychiatry*, **50**, 1542–1545.
Peacock, W.J., Arens, L.J. (1982) 'Selective posterior rhizotomy for the relief of spasticity in cerebral palsy.' *South African Medical Journal*, **62**, 119–124.
Penn, R.D. (1988) 'Intrathecal baclofen for severe spasticity.' *Annals of the New York Academy of Sciences*, **153**, 157–166.
Péterfi, T. (1927) 'Die Abhebung der Befruchtungsmembran bei Seeigeleiern.' *Archiv für Entwicklungs Mechanik der Organismen*, **112**, 660–695.
Pettigrew, J.B. (1873) *Animal Locomotion, or Walking, Swimming, and Flying*. London: King.
Peyton, A.J. (1986) 'Determination of the moment of inertia of limb segments by a simple method.' *Journal of Biomechanics*, **19**, 405–410.
Pflug, A.E., Aasheim, G.M., Foster, C., Martin, R.W. (1978) 'Prevention of post-anaesthesia shivering.' *Canadian Anaesthetists' Society Journal*, **25**, 43–49.
Proffit, W.R., Fields, H.W., Nixon, W.L. (1983) 'Occlusal forces in normal and long-faced adults.' *Journal of Dental Research*, **62**, 566–570.
Rack, P.M.H. (1985) 'Stretch reflexes in man: the significance of tendon compliance.' *In* Barnes, W.J.P., Gladden, M.H. (Eds) *Feedback and Motor Control in Invertebrates and Vertebrates*. London: Chapman and Hall.
Ralston, H.J., Libet, B. (1953) 'The question of tonus in skeletal muscle.' *American Journal of Physical Medicine*, **32**, 85–92.
Roberts, T.D.M. (1963) 'Rhythmic excitation of a stretch reflex, revealing (a) hysteresis and (b) a difference between the responses to pulling and to stretching.' *Quarterly Journal of Experimental Physiology*, **48**, 328–345.
—— (1967) *Neurophysiology of Postural Mechanisms*. London: Butterworths.
—— (1985) *Basic Skills of Horse Riding*. London: Allen.
Robinson, D.A. (1964) 'The mechanics of human saccadic eye movement.' *Journal of Physiology*, **174**, 245–264.
Robson, P. (1970) 'Shuffling, hitching, scooting or sliding: some observations in 30 otherwise normal children.' *Developmental Medicine and Child Neurology*, **12**, 608–617.
—— (1984) 'Prewalking locomotor movements and their use in predicting standing and walking.' *Child: Care, Health and Development*, **10**, 317–330.
Rodger, I.W., Bowman, W.C. (1983) 'Adrenoceptors in skeletal muscle.' *In* Kunos, G. (Ed.) *Adrenoceptors and Catecholamine Action. Part B*. Chichester: Wiley.
Rosett, J. (1924) 'The experimental production of rigidity, of abnormal involuntary movements and of abnormal states of consciousness in man.' *Brain*, **47**, 293–335.
Rosner, E. von (1956) 'Risus sardonicus.' *Forschungen und Fortschritte*, **30**, 103–105.
Rothwell, J.C., Traub, M.M., Day, B.L., Obeso, J.A., Thomas, P.K., Marsden, C.D. (1982) 'Manual motor performance in a deafferented man.' *Brain*, **105**, 515–542.
Rowe, D.W., Shapiro, J.R. (1989) 'Heritable disorders of structural proteins.' *In* Kelley, W.N., Harris, E.D., Ruddy, S., Sledge, C.B. (Eds.) *Textbook of Rheumatology*. Philadephia: W.B. Saunders.
Rubin, S. (1974) *Medieval English Medicine*. London: Abbot.
Saint-Hilaire, M.-H., Saint-Hilaire, J.-M., Granger, L. (1986) 'Jumping Frenchmen of Maine.'

Neurology, **36**, 1269–1271.
Sale, D., Quinlan, J., Marsh, E., McComas, A.J., Belanger, A.Y (1982) 'Influence of joint position on ankle plantarflexion in humans.' *Journal of Applied Physiology*, **52**, 1636–1642.
Salmons, S., Henriksson, J. (1981) 'The adaptive response of skeletal muscle to increased use.' *Muscle and Nerve*, **4**, 94–105.
Schaltenbrand, G. (1929) 'Muscle tone in man.' *Archives of Surgery*, **18**, 1874–1885.
Schieppati, M. (1987) 'The Hoffmann reflex: a means of assessing spinal reflex excitability and its descending control in man.' *Progress in Neurobiology*, **28**, 345–376.
Schivelbusch, W. (1980) *The Railway Journey*. (Translated by A. Hollo.) Oxford: Blackwell.
Schmidt, R.F. (1978) (Ed.) *Fundamentals of Neurophysiology*. New York: Springer.
Schmidt-Nielsen, K. (1984) *Scaling. Why is Animal Size so Important?* Cambridge: Cambridge University Press.
Sheldon, J.H. (1960) 'On the natural history of falls in old age.' *British Medical Journal*, **2**, 1685–1690.
Sherrington, C.S. (1919) 'Note on the history of the word "tonus" as a physiological term.' In *Contributions to Medical and Biological Research Dedicated to Sir William Osler. Vol. 1.* New York: Hoeber. pp. 261–268.
Simonson, E., Snowden, A., Keys, A., Brozek, J. (1949) 'Measurement of elastic properties of skeletal muscle in situ.' *Journal of Applied Physiology*, **1**, 512–525.
Smith, A.E., Martin, S., Garvey, P.H., Fenn, W.O. (1930) 'A dynamic method for measurement of muscle tonus in man.' *Journal of Clinical Investigation*, **8**, 597–622.
Smith, R.D. (1982) 'Paganini's hand.' *Arthritis and Rheumatism*, **25**, 1385–1386.
Soliman, M.G., Gillies, D.M.M. (1972) 'Muscular hyperactivity after general anaesthesia.' *Canadian Anaesthetists' Society Journal*, **19**, 529–535.
Springer, R. (1914) 'Untersuchungen über die Resistenz (die sogenannte Härte) menschlicher Muskeln.' *Zeitschrift für Biologie*, **63**, 201–222.
Stevenson, S., Lawson, J. (1986) *Masterpieces of Photography from the Riddell Collection*. Edinburgh: Scottish National Portrait Gallery.
Stone, S.L., Thomas, G., Zakian, V. (1965) 'The passive rotary characteristics of the dog's eye and its attachments.' *Journal of Physiology*, **181**, 337–349.
Styf, J., Korner, L. (1986) 'Microcapillary infusion technique for measurement of intramuscular pressure during exercise.' *Clinical Orthopaedics and Related Research*, **207**, 253–262.
—— —— Suurkula, M. (1987) 'Intramuscular pressure and muscle blood flow during exercise in chronic compartment syndrome.' *Journal of Bone and Joint Surgery*, **69B**, 301–305.
Swanson, S.A.V. (1963) *Engineering Dynamics*. London: English Universities Press.
Swift, J. (1838) *Voyages de Gulliver*. Paris: Furne.
Tauber, E.S., Coleman, R.M., Weitzman, E.D. (1977) 'Absence of tonic electromyographic activity during sleep in normal and spastic nonmimetic skeletal muscles in man.' *Annals of Neurology*, **2**, 66–68.
Thilmann, A.F., Fellows, S.J., Garms, E. (1990) 'Pathological stretch reflexes on the "good" side of hemiparetic patients.' *Journal of Neurology, Neurosurgery and Psychiatry*, **53**, 208–214.
—— —— Ross, H.F. (1991) 'Biomechanical changes at the ankle joint following stroke in man.' *Journal of Neurology, Neurosurgery and Psychiatry*, **54**, 134–139.
Thomasen, E. (1948) *Myotonia. Thomsen's Disease (Myotonia Congenita), Paramyotonia, and Dystrophia Mystonica*. Denmark: Univ. Aarhus.
Thompson, D.T., Wright, V., Dowson, D. (1978) 'A new form of knee arthrograph for the study of stiffness.' *Engineering in Medicine*, **7**, 84–91.
Thompson, D.W. (1942) *On Growth and Form*. Cambridge: Cambridge University Press.
Thurston, G.B., Greiling, H. (1978) 'Viscoelastic properties of pathological synovial fluids for a wide range of oscillatory shear rates and frequencies.' *Rheological Acta*, **17**, 433–445.
Tsementzis, A., Gillingham, F.J., Gordon, A., Lakie, M. (1980) 'Two methods of measuring muscle tone applied in patients with decerebrate rigidity.' *Journal of Neurology, Neurosurgery and Psychiatry*, **43**, 25–36.
Walker, J. (1982) 'The essence of ballet maneuvers in physics.' *Scientific American*, **246**, 118–126.
Waller, A.D. (1896) *An Introduction to Human Physiology*. 3rd Edn. London: Longmans Green.
Walsh, E.G. (Ed.) (1963) *Cerebellum Posture and Cerebral Palsy*. Little Club Clinics, No. 3. London: Heinemann. pp. 31–37.

—— (1966) 'Head movements during rail travel.' *Biomedical Engineering*, **1**, 402–407.
—— (1968a) 'Model of load-seeking reflex—a "Pugnatron".' *Journal of Physiology*, **197**, 57–59P.
—— (1968b) 'Oscillations of the head, trunk and limbs induced by mechanical means.' *Journal of Physiology*, **198**, 69–72P.
—— (1970a) 'Tremor of the wrist induced by positive velocity feedback.' *Journal of Physiology*, **207**, 16–17P.
—— (1970b) 'Oscillatory transients in the response to an abrupt change of force at the wrist in Parkinsonism.' *Journal of Physiology*, **209**, 33P.
—— (1970c) 'Balancing movements of standing man.' *In* Simpson, D.C. (Ed.) *Modern Trends in Biomechanics—1.* London: Butterworths.
—— (1971) 'Ankle clonus—an autonomous central pacemaker?' *Journal of Physiology*, **212**, 38–39P.
—— (1973a) 'Motion at the wrist induced by rhythmic forces.' *Journal of Physiology*, **230**, 44–45P.
—— (1973b) 'Motion at the ankle induced by random forces—frequency analysis.' *Journal of Physiology*, **234**, 100–101P.
—— (1973c) 'Standing man, slow rhythmic tilting, importance of vision.' *Agressologie*, **14C**, 79–85.
—— (1974) '"Pugnatron"-like reaction in a patient with familial dystonia: torque induced motion analysis.' *Journal of Neurology, Neurosurgery and Psychiatry*, **37**, 559–565.
—— (1975a) 'Resonance at the wrist—a "jump" effect.' *Journal of Physiology*, **245**, 69–70P.
—— (1975b) 'Jump effect at the wrist—model using light or non-linear damping.' *Journal of Physiology*, **248**, 19–20P.
—— (1976a) 'Shortening reactions in the human forearm.' *Journal of Physiology*, **256**, 116–117P.
—— (1976b) 'Stretch reflexes in forearm muscles.' *Journal of Physiology*, **263**, 263P.
—— (1976c) 'Clonus: beats provoked by the application of a rhythmic force.' *Journal of Neurology, Neurosurgery and Psychiatry*, **39**, 266–274.
—— (1977) 'Persistence of stretch reflexes following cerebellar ablation—and a resonance theory of cerebellar function.' *In* Rose, F.C. (Ed.) *Physiological Aspects of Clinical Neurology.* Oxford: Blackwell.
—— (1979) 'Beats produced between a rhythmic applied force and the resting tremor of Parkinsonism.' *Journal of Neurology, Neurosurgery and Psychiatry*, **42**, 89–94.
—— Lakie, M., Wright, G.W., Tsementzis, S.A. (1980) 'Measurements of muscle tone using printed motors as torque generators.' *Engineering in Medicine*, **9**, 167–171.
—— Wright, G.W. (1987a) 'Inertia, resonant frequency, stiffness and kinetic energy of the human forearm.' *Quarterly Journal of Experimental Physiology*, **72**, 161–170.
—— —— (1987b) 'Patellar clonus: an autonomous central generator.' *Journal of Neurology, Neurosurgery and Psychiatry*, **50**, 1225–1227.
—— —— (1987c) 'Resonance at the human hip.' *Journal of Physiology*, **392**, 4P.
—— —— (1988) 'Postural thixotropy at the human hip.' *Quarterly Journal of Experimental Physiology*, **73**, 369–377.
—— —— Powers, N., Nuki, G., Lakie, M. (1989) 'Biodynamics of the wrist in rheumatoid arthritis—the enigma of stiffness.' *Proceedings of the Institution of Mechanical Engineers*, **203**, 197–201.
—— Brown, K., Wright, G.W. (1990a) 'Resonant frequency of the human ankle in juvenile hemiplegia.' *Journal of Physiology*, **429**, 31P.
—— Wright, G.W., Brown, K., Bell, E. (1990b) 'Biodynamics of the ankle in spastic children—effect of chronic stretching of the calf musculature.' *Experimental Physiology*, **75**, 423–425.
——Lambert, M., Wright, G. W., Powers, N., Nuki, G. (1991) 'Resonant frequency at the wrist in hypermobile women.' *Experimental Physiology*, **76**, 271–275.
—— Wright, G.W., Lin, J.-P. (1992) 'Differences in the mechanogram of the calf musculature between male and female medical students.' *Pericirculated communication for St Andrew's meeting of the Physiological Society, 11th–13th June.*
Wartenberg, R. (1951) 'Pendulousness of the legs as a diagnostic test.' *Neurology*, **1**, 18–24.
Webster, D.D. (1964) 'The dynamic quantitation of spasticity with automated integrals of passive motion resistance.' *Clinical Pharmacology and Therapeutics*, **5**, 900–908.
—— (1966) 'Rigidity in extrapyramidal disease.' *Journal of Neurosurgery*, **24** (Suppl. Part 2), 299–309.
Wiegner, A.W. (1987) 'Mechanism of thixotropic behaviour at relaxed joints in the rat.' *Journal of Applied Physiology*, **62**, 1615–1621.

Westphal, G. (1880) 'Über eine Art paradoxer Muskelcontraction.' *Archiv für Psychiatrie und Nervenkrankheiten*, **10**, 243–248.

Whitlock, J.A. (1990) 'Neurophysiology of spasticity.' *In* Glenn, M.B., Whyte, J. (Eds.) *The Practical Management of Spasticity in Children and Adults.* Philadelphia: Lea and Febiger.

Williams, F.C., Uttley, A.M. (1946) 'The velodyne.' *Journal of the Institution of Electrical Engineers*, **93**, 1256–1274.

Winstanley, M. (1986) 'The tender touch of rigor mortis.' *New Scientist*, **110**, 36–39.

Wisner, A., Donnadieu, A., Berthoz, A. (1964) 'A biomechanical model of man for the study of vehicles seat and suspension.' *International Journal of Production Research*, **3**, 285–315.

Woods, A.G. (1967) 'Human response to low frequency sinusoidal and random vibration.' *Aircraft Engineering*, **39**, 6–14.

Wright, V., Johns, R.J. (1960) 'Physical factors concerned with the stiffness of normal and diseased joints.' *Bulletin of the Johns Hopkins Hospital*, **106**, 215–231.

Yemm, R., Nordstrom, S.H. (1974) 'Forces developed by tissue elasticity as a determinant of mandibular resting posture in the rat.' *Archives of Oral Biology*, **19**, 347–351.

Yung, P., Unsworth, A., Haslock, I. (1984) 'Measurement of stiffness in the metacarpophalangeal joint: circadian variation.' *Clinical Physics and Physiological Measurement*, **5**, 57–65.

INDEX

A

Acceleration
  equation, 37
  rate, 67
Accelerometer, 43
Action and super-reaction, 130–1
Active touch, 128 (fig.)
Adrenaline and physiological tremor, 97
Aircraft, military, vibration, 174
Akathisia, 17
Alexander method, 20
Alfentanil, 159
Allometry, 75–7
Alpha rigidity, 183
Alternating torques, **118**
Anabolic steroids, 5
Anaesthesia
  effects on wrist, 54 (figs)–5
  shivering after, 160
André-Thomas, **8–11**
Andry, N., 19–20 (fig.)
Ankle
  clonus, 146 (fig.)–7, 186
  reaction, 131 (fig.)
  resonance in juvenile hemiplegia, 151 (fig.)
  vibrational 'noise' assessment, 180 (fig.)–1
Ankle jerk, recording method, 28 (fig.)
Anxiety, chronic, 1
Appleton, Sir Edward, 140
Arm, extensibility tests, 9 (fig.)
Ashworth scale, **8** (table)
Asymmetry, assessment, 11
Ataximeters, 19
Athetosis, after hemiplegia, 182 (fig.)–3, 184 (fig.)
Athlete, muscle swelling, 23
Atlas, C., 3
Audiograms, 15 (fig.)

B

Benign hypermobility syndrome, 105
Bhopal gas tragedy, 126
Biceps, pressure, 23
Biofeedback, 56
Biological survival value, 76 (fig.)–7
Bipennate fibre arrangement, 11
Birds, wings as mechanical oscillators, 183–4
Body
  size effects, 75–7
  vibrations of low frequency, 169–70
Borelli, G. A., 11
Boxing, momentum, 67
Brain, shearing by rotary acceleration, 172 (fig.)

C

Cadaveric spasm, 165
Cardiac surgery, cooling for, 80
Carriage
  horse-drawn, 174
  laboratory induced motion, 178–9 (fig.)
Catalepsy, **161**
Cataplexy, **116**
Catatonia, **161**
Cerebellum
  as adaptive control system, 190
  function studies, 182 (figs)–3
  in muscle tone, **116–17**
  lesions, **183**
Cervical spondylosis, **146 (fig.)**
Chirps,
  48, 51 (fig.)
  in Parkinsonism, 135 (fig.)
  for resonant frequencies, 70
  in spastic wrist, 142 (fig.)
Clonus
  ankle, **146 (fig.)–7**, 186
  description, 147
  interference, 148 (fig.)
  patellar, **150**
  wrist, **150**
Clumsiness, tests for, 189 (table)
CNS, changes associated with learning, 188
Co-contraction, unmodulated, 137 (figs)–42
Cogwheel rigidity, **136–7 (fig.)**
Cogwheeling, **136 (fig.)–7 (fig.)**
Cold adiadochokinesia, 83–4 (fig.)
Compartment syndrome
  acute, 23
  chronic, 23
Concussion, **170**
Connective tissue hereditary disorders, 105
Contact lens, 90–1
Contracting muscles, sounds generated, 14–16
Contraction time, 13
Cooling, changes induced, 80–3
Crabs, thixotropic effects, 89–90 (fig.)
Cramp, **156–9**
Cubit, 73 (fig.)
Cursorial animals, adaptations, 168–9

D

Damping, **57–8**
  measurement, 115 (fig.)
  in resonant frequency, 144 (figs)–5
Damping coefficient (C), 61
Darwin, C., 1–2 (fig.)

de Ajuriaguerra, 8–11
Decerebrate rigidity, **145**
Delpech, J., 19–20
Delpech's method, 16
Dentatomy in cerebral palsy, 183
Diazepam, 31
Displacement, predetermined rate *see* Stretching
Dorsiflexing bias, 147 (fig.)
   slow variation, 147 (fig.)
   with inertia, 148 (fig.)
Drag, 35
Driver, body resonant properties, 180 (fig.)
Driving force, 78
Drop attacks, **116**
Duchesne, 177
Dynamic magnifier, 57
Dystrophia myotonica, 83

E
Ear drum, voluntary deafness, 14–15 (fig.)
Ears, muscular sounds, 14–16
Elastic loading, **185–6**
Elbow, biodynamics study, 72 (fig.)
Electromyogram (EMG), 1
   manual displacement measurement, 25
   in maximal stiffening, 121–2 (fig.)
   needle-type, 55
   in tension, 121 (fig.)
   wrist, 48
Emotions, effects on muscles, 1–2 (fig.)
Engineers' illness, 177
Epilepsy, 161 (fig.)
Etymology, **5–7**
Excitation-contraction coupling, 159
Extensibility, tests, 9 (fig.)–10 (fig.), 11
Eye
   biodynamics, **90–4**
   low-frequency alternating torque, 92 (fig.), 94
   low-frequency ramped torque, 92 (fig.), 94
   movement recording apparatus, 91 (fig.)
   torque pulses, 92 (fig.), 94
   vertical movements, 93 (fig.)
Eyeball and gravity, 102

F
F response, 29
Facial palsy, 1
Faraday disc, 45 (fig.)
Fasciotomy, 23
Feedback
   negative, 59 (fig.)
   negative velocity, 59
      measurement methods, 71–5
   positive velocity, 59, 63, 70 (fig.)
      measurement methods, 71–5
      ramped, 63–4 (fig.)
   stroke therapy, 64 (fig.)–6
   using varied torque levels, 63–4 (fig.)
Finger
   hypotonia, 103–4 (fig.)
   thixotropy measurements, 98–9
Flexor pollicis longus, tendon, 167 (fig.)
Floppy infants, 103
Force, defined variation *see* Pulling
Force transmission, 166–7
Forearm
   cooling apparatus, 82 (fig.)
   cooling effect on wrist compliance, 119 (fig.)
   experimental cooling, 82 (figs)–3
   inertia relationship, 73 (fig.)
   oscillations slowed by inertia, 149 (fig.)–50
   stiffness measurement, 69–71
Fourier analysis, 100 (fig.)
Frequency analysis, 99

G
Galileo, 78
Gamma rigidity, 183
Gastrocnemius, 167 (fig.)
Gear ratio, **167–9**
Goniometry, electric, 152 (table)
Gowers' sign, 4 (fig.)–5
Grasp reflexes, **160–1**

H
H reflex, 26–9
   in Holmes-Adie syndrome, 123
   pathway, 26 (fig.)
   vibration inhibition, 114 (fig.)
Hand
   inertia measurement, 69–70 (fig.)
   physiological tremor, 95 (fig.)
   postural tremor, 96 (fig.)–7
   suspended, effect of taps, 97–8 (figs), 99 (figs)–100 (fig.), 101
Hanging hand tremorograph, 93 (fig.)–8
Harvey, W., 23
Head
   load-carrying, **170**
   mechanics of injuries, 171–2 (fig.)
Hemiplegia
   juvenile, 149 (fig.)
   tension in muscle movement, 24
   wrist resonance, **142–3 (figs), 144 (fig.)–5**
Hip,
   biodynamics, **85–9**
   thixotropic behaviour, 87 (fig.), 89
Holmes-Adie syndrome, stretch reflexes, **123** (fig.)
Hooke, Robert, 172
Hooke's law, 52, 83

Horse
    movements, **172–4**
        head, **172**
        trotting, **173 (figs)**
Hunter, J., 3, 166
Huntington's chorea, 190
Hyperextensibility, 10 (fig.)–11
Hypermobility, **105–7**
    tests, 106 (fig.)–7
Hyperpyrexia, malignant, 159–60
Hypertonia
    and anaesthesia, 159–60
    episodic, **161–2**
    skeletal muscle, 31
Hypothermia, 80
Hypotonia, **103–17**
    vibration-induced, 113–14
Hysteresis loop, 35–6 (fig.)

I

Impedance component, 41
Inertia, **35**
    estimation, 37
    moment, **67**
        estimates, 68–9
    quick release measurement, 68 (fig.)
Inertial loading, **185**
Infant
    creeping, 193–4
    floppy, 103
    injuries caused by shaking, **170**
    neurological examination, 9–11
    'shuffling', 193–4
Injuries caused by shaking infants, **170**
Intramuscular pressure, measurements, 23
Invertebrate muscular locking, 12
Invertebrates, thixotropy, 89–90
Isokinetic stretching, 38–40

J

J-receptor
    activity, muscle tone reduction, **126–7**
    reflex effects on muscle, 127
    stimulation, 127 (fig.)
Jaw
    biomechanics investigation, 124 (fig.)–5
    damping, 126
    forces applied, 124 (fig.)
    movement, 124 (fig.)–5
    musculature tone, **125–6**
    positive velocity feedback, 62 (fig.)
    power-weight ratio, 126
Jaw jerk, 125
Joint
    laxity, 105
    mobilisation, 66 (fig.)
    motion control, 107
    resonant frequencies comparison, 181 (table)–3
Jumping Frenchmen of Maine, 161–2

K

Karate, techniques, 67
Ketamine, 159
Knee
    measurements after stroke, 109 (fig.)
    normal variability, 110 (table)
    variability after stroke, 110 (table)

L

L-DOPA, evaluation, 133
Lead pipe rigidity, 31
Leg
    changes in stroke, 108 (fig.)
    motion measurement in stroke, 108 (fig.)
    swing measurement, 86 (figs)
Limbs
    double-jointed, 105
    movement, muscle control, 188 (fig.)–9
    muscle proportion, 5
    resistance to passive bending, 53
    resonances, 43
    segment coupled movements, 63–5
Lobeline, 127
Locomotion, **169**
Loosening, caused by disturbance, 78–9 (fig.), 80
Lotus position, 104 (fig.)–5
Lysozymes, 165

M

McArdle's disease, 158
Magnification factor, 57
Marey, E. J., 173
Mental activity and frontalis muscle, 1
Midazolam, effect in multiple sclerosis, 32 (fig.)
Mitchell, Weir, 162
Mobility, calculation, 63, 75
Mollusc, vibration effects, 113
Morphine, 159
Mosso, A., 29, 78, 156, 187
Motion, resistance, 34 (fig.)–7
Motor units, **13**
Movements
    19th-century recording method, 168 (fig.)
    voluntary into automatic, 187
Multipennate fibre arrangement, 11
Muscle consistency, test, 9
Muscle fatigue, 16–17 (fig.)
Muscle fibres, **11–13**
    different types, 13–14
    distribution, 12 (fig.)

Muscle response, pathway, 26 (fig.), 28
Muscle strength, measurement, 75
Muscle tone
  clinical assessment, 7 (fig.)–11
  description, 6–7, 53
  electrical measurement, 26 (figs)–9
  gravity-driven measurement, 29–33
  in human refrigeration, 80–1
  increase after anaesthetic, 159
  manual displacement measurements, 24–5 (fig.)
  test, 7 (fig.)
Muscles
  development, **2–5**
  exercise changes, 63
  'full action', 22
  'irrelevant' activity, 57 (fig.)
  large, active tone, 16–17
  'short action', 22
  strength, **75**
  thixotropic effects, 83–4
Musculo-tonic action, 53
Musculoskeletal complaints of hypermobility, 105
Myasthenia gravis, 1–2
Myoglobin, 13
Myosclerometer, 23–4 (fig.)
Myositis ossificans, **163**
Myotonia, 81–3
Myotonia congenita, 81
Myotonograph, 28 (fig.)

N

Neck stiffening, 171
Negative velocity feedback, **59**
Nerve injuries, **162**
Neural influences in torque levels, **52–7**
Neuromuscular blocking drugs, thixotropic effects, 80

O

Occupational cramps, 156–7
Ocydromes, 168 (fig.)
Orthopaedics, 19
Oscillations, induced by locomotion, 168 (figs)–70
Oscillators, rhythmic input effects, 140

P

Paganini, N., 106
Parameters in relaxed adults, 74 (table)
Paramyotonia, 83
Paraplegia
  muscle state, **107–8**, 113
  spasm in, 155 (fig.)
Paraplegia *see also* Stroke

Parkinsonism, **133–42**
  activation, 138
  dystonic stage, 137 (fig.)
  muscle tone measurement, 43–52
  resonant frequencies of wrist, 134 (fig.)
  steady tremor, 141 (fig.)
  tension in muscle movement, 24
  test, 8
  Wartenberg's test, 30 (fig.)
Passivity, test, 11
Patellar clonus, 150–1 (fig.)
Patellar reflex, vibration effect, 114 (fig.)
Pennate fibres, 22
Pes equinus, **150–3**
  serial casting, 152–3
    ankle resonant frequencies, 152 (fig.)
Phalanx, terminal, disconnected, 103
Phenol, effect in multiple sclerosis, 32 (fig.)–3
Phenothiazines, 17
Physiotherapy
  automated, 65–6 (fig.)
  significance of thixotropy, 101
Polyneuritis, 5
Position sense, 19
Positive velocity feedback, **59**
Post-infection fatigue syndrome, 75
Postural activity, pugnatron-like, 128–32
Postural control, significance of thixotropy, 101–2
Postural sway, 18 (figs)–19
Postural system, 10 (fig.)–11
Postural tone, 16–17
Posture, resonance theory, 186–90
Potentiometer, 73
Printed motors, **45–7** (fig.)
  armature, 46 (fig.)
  to apply wrist torque, 111 (fig.)
Ptosis, 2 (fig.)
Pugnatron, **128 (fig.)–9**
Pulvinar lesion, 131 (fig.)–2
Purkinje, J., 178

Q

Quality factor Q, 57
  formula, 75

R

Rabies, 154 (fig.)–5
Racing cyclists, neuromuscular aberrations, 157 (fig.)–158 (fig.)
Railway carriage
  early, shaking, 174–5 (fig.), 176–178 (fig.)
  floor acceleration, 177 (fig.)
  invalid, 176 (fig.)
  passenger's head movements, 177 (fig.)
Railway nystagmus, 178 (fig.)

Refrigeration
    meat, 164–5
    muscle tone, 80–1
Relaxation index, 31
Relaxation therapy, 56
Relaxation time, 13–14
Resonance
    curves in anaesthesia, 54 (fig.)
    peak velocity, 57–8 (fig.)
    principles, 50 (fig.)
Resonant frequency, **48–9** (figs), 51 (fig.)
    comparison of joints, 181 (table)–3
    inertia measurement, 69–70 (fig.), 71 (fig.)
    leg swing, 87 (figs)
    in Parkinsonism, 133–4 (fig.), 135 (fig.)
    stiffness effects, 72 (fig.)
Rheumatoid arthritis
    resonant frequency and peak torque relationship, 115 (fig.)
    stiffness problem, **115–16**
Rhythmic forces, for muscle tone assessment, 43–4 (figs)
Rigidometer, 40
Rigor, **163–5**
Rigor mortis, rabbit 164 (fig.)
Royal Hospital for Sick Children, Edinburgh, 170
Running, tendon action, 167

S
Salbutamol and hand tremor, 96 (figs)
Sarcomere, 22
Scale, **75**
Selective posterior rhizotomy, 153
Sherrington, C., 11
Shivering, **160**
Short-range elastic component, 83
Shortening reactions, 120 (fig.)
    in forearm, 121 (fig.)
    in Parkinsonism, 135 (fig.)–6
Shoulder, load-carrying, 171 (fig.)
Shrapnell's membrane, 15
Sinusoidal motion, 40–1
Sinusoidal torques, eye motion, 92–3 (figs), 94
Skeletal adaptation, 3
Skeletal deformities, tone in, 19
Soleus, 167 (fig.)
Spasms, **154–6**
Spastic, origin, 6
Spastic paraparesis, anaesthetic injection effect, 33
Spasticity, **142–54**
    after stroke, 107
    hypertonus assessment, 40
Speed of movement, 40
Squatting facet, 105

Stabilisation, 63
Step function use, 118
Stiff baby syndrome, **161**
Stiff man syndrome, **161**
Stiffening
    with small forces, 78–9 (fig.)
    voluntary, 120–3
    and wrist resonance, 49 (fig.)
Stiffness, 33 (fig.)–5
    elastic, measurements, 41 (fig.)–2
    estimation, 36 (fig.)–7
    inertial, 34 (fig.)–5
    measurement, 72 (fig.), 114–15 (figs)
Stirring force, 78, 81 (fig.)
Strain *see* Movement
Stress *see* Torque
Stress-relaxation, 25
Stretch reflex, 27, 40
    discovery, 55
    increase after decerebellation, 182 (fig.)–3
    operation, 120–3
    phasic, 11
    reduction by phenyldiguanide, 127 (fig.)
    in voluntary stiffening, 122 (fig.)
Stretching
    rhythmic isokinetic methods, 38–9 (fig.)
    sinusoidal methods, 40–1
    unidirectional isokinetic methods, 38–9 (fig.), 40
Stroke
    leg changes, 108 (fig.)
    leg motion measurement, 108 (fig.)
    positive velocity feedback therapy, 64 (fig.)–6 (fig.)
    Wartenberg's test, 30 (fig.)
Stroke *see also* Paraplegia
Swinging of arms, 169

T
Tabes dorsalis, muscle weakness, 4 (fig.)
Tachometer, 73
Tank method for forearm cooling, 119
Telephoning, wrist spasticity activation, 145 (fig.)
Tendon jerks, 25–7
Tendon retraction, 22
Tendons, **166–7**
Tenotomy, 153
Tension in training, 3
Tetanus, **162–3** (fig.)
Tetany, **162**
Thalamo-pallidal stereotaxy, wrist flexibility, 133–4 (fig.)
Thixotropic effects, 41
    in body parts, 88 (fig.)–9
Thixotropic memory, measurement, 85 (fig.)